STEM
Education
for High-Ability Learners
Designing and Implementing Programming

STEM Education
for High-Ability Learners
Designing and Implementing Programming

Edited by
Bronwyn MacFarlane, Ph.D.

Copublished With the

NATIONAL ASSOCIATION FOR
Gifted Children

Routledge
Taylor & Francis Group

NEW YORK AND LONDON

Library of Congress Cataloging-in-Publication Data

STEM education for high-ability learners : designing and implementing programming / edited by Bronwyn MacFarlane.
   pages cm
 ISBN 978-1-61821-432-4 (pbk.)
1. Science--Study and teaching--United States. 2. Technology--Study and teaching--United States. 3. Engineering--Study and teaching--United States. 4. Mathematics--Study and teaching--United States. 5. Gifted children--Education--Curricula--United States. 6. Curriculum planning--United States. I. MacFarlane, Bronwyn, 1974- editor of compilation.
 Q183.3.A1S728 2015
 507.1'273--dc23
               2015006440

First published in 2016 by Prufrock Press Inc.

Published in 2021 by Routledge
605 Third Avenue, New York, NY 10017
2 Park Square, Milton Park, Abingdon, Oxon OX14 4RN

Routledge is an imprint of the Taylor & Francis Group, an informa business

ISBN: 9781032142142 (hbk)
ISBN: 9781618214324 (pbk)

DOI: 10.4324/9781003238218

Table of Contents

Dedication

To my supportive family, Beverly and Greg, and in memory
of Ted and Brian, who continue to inspire.

Introduction

Nearly two-thirds of the jobs in the 21st century economy are high-skill positions (Gordon, 2009), and the American workforce of 2009 had fewer than half the number of qualified candidates needed to fill those positions. This lack of supply forces employers to choose among outsourcing jobs, importing skilled employees, or relocating operations to overseas markets, where there exists a growing supply of skilled workers. Gordon predicted that by 2020, three-quarters of the job market will require high-skill workers. Developing talented technical professionals to problem solve, create new innovations, and fill high-skill positions is the perpetual work of the educational system. This book provides a comprehensive resource for educators designing and implementing STEM education programs by providing a research-based discussion of each critical component for inclusion in a planned, coherent, and quality-minded sequenced system—specific to STEM education—for advanced learners.

Knowledge Base: Differentiated STEM Learning for High-Ability Learners

This text intends to provide discussion about a continuum of services for the talent development of promising performers in the STEM fields. Based on the theory and research of what works in gifted education, differentiation, and STEM educational projects, this text has been developed for creating a coherent system of learning that maximizes available resources. Just as gifted students need differentiated school-based curriculum, advanced students with technical talent need their STEM learning experiences to be differentiated for their individual learning needs as well. This text provides a complete set of chapters addressing the specific issues related to supporting advanced learning in the STEM areas. Readers will gain professional knowledge, understanding, reflection, and reform ideas for supporting talent trajectories in the STEM fields. Supportive areas of STEM study, including science, technology, engineering, mathematics, as well as the arts, creativity, and second language study, are incorporated to either stand alone or fit together as critical pieces of a well-balanced learning experience that translates for many professional opportunities. Chapter content presents information ranging across STEM content area planning, pedagogy, standards, resources, and assessment.

Each chapter provides a discussion regarding research findings, best practices in supporting talent in STEM education, and questions for reflection and service improvement. Chapter material links to the major themes in gifted education, such as interdisciplinary learning, curriculum rigor, differentiation, and affective development issues. Connections to the larger education field are also clear with specific discussions about topics, such as the Next Generation Science Standards, the Common Core State Standards, the Partnership for 21st Century Skills, assessment, and professional training considerations.

STEM Education for High-Ability Learners: Designing and Implementing Programming provides a professional publication focused on the rigorous articulation of STEM educational programming to develop STEM talent among learners with high abilities. Within this edited textbook, each chapter provides a cutting-edge discussion of best practices for delivering STEM education by experts in the field. Hopefully, STEM educators will find these chapters beneficial when considering how to support and develop the talent of advanced learners in each area discussed. As the chapters reflect, educators should not only focus upon exposing students to STEM content, but also facilitate high performance outcomes across the STEM disciplines through use of best practices in education.

Chapter Authors

The contributing authors of the individual chapters represent a range of experience levels, from senior scholars and administrators to promising, early career researchers. Invited authors include expert professionals active in the field of gifted education and scholars working in related fields whose input provides a valuable perspective to the diverse elements for inclusion in a comprehensive STEM education program.

Use of the Book

The intended audience for this edited volume includes STEM educators, school administrators, department coordinators, central office personnel, graduate students, university personnel, gifted program coordinators, and teachers. Parents who home school children and coordinators of STEM educational initiatives offered by external organizations are also within the primary audience. This volume may be put to several uses, including professional development reading, college course reading, and training in STEM education. This text will help educators reflect and plan for differentiated STEM services and instruction for high-ability students. This text can be used in university courses for discussions about program delivery, instructional design, professional development, and systems of support, as well as school-based professional development. Possible course adoptions include: Methods for STEM Education, Curriculum and Instruction for the Gifted, Differentiating in the Regular Classroom, Current Issues and Research, Educational Policy, School Reform, and other related courses.

The Organization of the Book

The book is delineated into two threaded parts in which chapters have been placed accordingly. Part I focuses on foundational theories, research, and best practices in STEM education, and chapters provide detailed information about the distinct core components of a comprehensive STEM program differentiated for advanced learners. Part II focuses on applications to school-based practice in gifted education and presents important information relevant to the infrastructural supports for a comprehensive STEM program differentiated for high-ability learners.

Each chapter follows a common general outline intended to provide continuity throughout the book and includes an introduction of the chapter topic, reviews relevant empirical literature and research findings of major STEM themes, and highlights practices in schools and in gifted education. Historical precedents and guiding principles are discussed and lend perspective to understanding new trends and directions in STEM education and connections to be cultivated between general and gifted education. Contributing authors also discuss unresolved issues and questions recommended for examination. Following the research discussion, many chapters also provide a practical discussion with best practices in serving gifted learners in STEM domains at various levels. Where relevant, connections to the standards are provided and practical applications are discussed for advanced curriculum delivery (e.g., sample activities, lessons, examples of assessments or programs, etc). The implications for teacher preparation of gifted learners relevant to specific topics are also discussed. Each chapter also includes excerpts from the text selected by the editor, which are highlighted in gray sidebars for emphasis. Finally, each chapter concludes with a set of questions to be used in concert with professional learning activities for stimulating critical reflection and discussion among the reading audience.

Chapter Orientation and Contents

In the first chapter, Dr. Julia L. Roberts provides insight about state residential STEM schools with a focus on mathematics and science. Fifteen states have a state residential school that highlights STEM subjects and provide opportunities for talented students to study at advanced levels with academic peers. Such schools allow students to accelerate their learning, especially in STEM subjects. In particular, Dr. Roberts provides a case study detailing The Carol Martin Gatton Academy of Mathematics and Science in Kentucky as an example of a state residential school that is located on the campus of Western Kentucky University (WKU).

In the second chapter, Dr. Debbie Dailey discusses elementary science curriculum and instruction for developing scientific thinking and habits among young learners. She presents research supporting engaging science practices, the use of various instructional strategies and teaching practices, ways to improve science in the elementary grades, and teacher science and pedagogical skills.

In the third chapter, Dr. Steve Coxon presents secondary level science curriculum and instructional methods. Key elements of secondary science curriculum are summarized; project- and problem-based learning are discussed as ideal methods for the secondary science classroom; and programming options including science-related advanced coursework, special schools, and extracurricular programs are discussed.

In the fourth chapter, by Drs. Angela and Brian Housand, the "T" in STEM (technology) is characterized as the entire system of people and organizations, knowledge, processes, and devices that go into using and creating technological artifacts. The authors provide clarity about technology education and educational technology and the differences that should be understood for effective use with learners in preparing students for technical jobs that do not yet exist using technologies that have yet to be imagined.

In the fifth chapter, by Drs. Debbie Dailey and Alicia Cotabish, engineering education and applied science is examined to cultivate students' understanding and problem-solving abilities. By providing students with applied experiences in both science and engineering practices, students will be more prepared to solve societal and environmental problems. Similarities and differences between problem-based and project-based learning are discussed, as well as how to integrate engineering with applied activities into science curriculum using standards.

In the sixth chapter, Drs. Scott Chamberlin and Eric Mann present elementary mathematics curriculum and instruction for advanced learners in primary school settings. The chapter is comprised of two sections. First, desirable characteristics of model curricula in mathematics education are discussed. Subsequently, four curricula that meet such characteristics are discussed. The chapter concludes with a recommendation of the *ideal* mathematics curriculum for STEM development of gifted learners.

In the seventh chapter, by Drs. Eric Mann and Scott Chamberlin, mathematics at the secondary level is discussed and readers are provided with specific discussion of the learning expectations associated with the Common Core State Standards in respect to the needs of advanced learners in upper grades. The authors provide options and recommendations for differentiating and enriching secondary mathematics curriculum to better meet the needs of talented students.

In the eighth chapter, Dr. Amy Sedivy-Benton, Ms. Heather A. Olvey, and Dr. James P. Van Haneghan share their understanding about assessing aptitude and achievement in STEM teaching and learning. They first explore methods for identifying STEM potential and aptitude and then discuss assessment practices that can be used to facilitate and measure the development of interest and achievement in STEM areas.

In the ninth chapter, Dr. Bronwyn MacFarlane reviews infrastructural elements essential for inclusion in STEM educational programming. A set of considerations and recommendations associated with planning programmatic and curricular design is provided, along with a discussion of these elements in application with advanced learners. STEM education is an interdisciplinary approach to leaning where rigorous academic concepts are coupled with real-world lessons. As educators plan and review STEM programs in local schools, this chapter provides an aerial view of programmatic, curricular, and instructional features to be included in cohesive STEM educational services.

In the 10th chapter, Dr. Alicia Cotabish connects the Common Core State Standards, the Next Generation Science Standards, and the Gifted Programming Standards with STEM curriculum for advanced learners. The articulation of multiple sets of standards calls for teachers to integrate objectives across standards and ultimately create a plan for elevated learning in all subject areas. By explaining strategies to plan and implement the new standards, this chapter presents a guide for creating cross-disciplinary content and integrating the standards.

In the 11th chapter, Dr. Ann Robinson, Ms. Kristy Kidd, and Dr. Mary Christine Deitz present the use of biography for building STEM understanding among talented learners. The history, rationale, and implementation of biography as a curricular strategy for understanding the real-world practices of scientists and engineers are explored in this chapter. Action research and qualitative studies from general and gifted education literature on incorporating STEM biographies are briefly reviewed. Because the Next Generation Science Standards are replete with opportunities to enrich STEM content and practices through biography, examples from children's biographies, the *Blueprints for Biography* curricular guides designed to accompany them, and lessons learned from implementing STEM biographies with talented learners are included.

In Chapter 12, Dr. Bronwyn MacFarlane provides a rationale for integrating second language study into STEM curriculum to prepare advanced STEM innovators for global opportunities. This chapter discusses the global nature of social interactions and why STEM programming should expand the curricular focus to include a range of language options, including German, Spanish, Chinese, and Arabic. By applying world language study within meaningful STEM project-based learning scenarios, students will be provided with opportunities to develop sophisticated critical skills sets to open a wide range of diverse opportunities across the STEM fields.

In Chapter 13, Dr. Rachelle Miller writes about integrating the arts and creativity in STEM education essential for architecture and design. She reviews empirical research and suggestions about gifted individuals producing innovation, the benefits of arts integration, teacher attitudes about arts integration,

STEM integration of arts curriculum, and suggests integrating the arts with STEAM. The art integration examples relate to one or more of the following art subjects included in the National Standards for Arts Education: dance, literary arts, media arts, music, theater, and visual arts.

In Chapter 14, Dr. Barbara Kerr and Mr. J. D. Wright deliver an academic discussion about gender differences research, the neuromythologies associated with gender, and the effects upon the STEM talent trajectories of gifted girls and boys. This chapter explores the history of studies about gender differences and the newest frontier of sex differences research in the field of neuroscience. Flaws in research design and researchers' interpretation of results are examined, as well as the ways in which the media exaggerate and misinform the public. Finally, this chapter shows how the treatment of sex differences in psychometric and brain studies undermine the STEM talent development of gifted boys and girls through stereotype threat, gendered education, inappropriate career guidance, and lack of support for nontraditional paths to fulfillment of potential.

In Chapter 15, Dr. Judy Stewart, Dr. Christopher Gareis, and Ms. Caroline Martin present a detailed picture about designing and developing an exceptional living and learning environment for high-ability students and describe specific information about a statewide STEM initiative. The authors share the seminal experiences and impactful accomplishments that have informed the design and development of the Virginia STEAM Academy. They outline the key steps to bringing the full scope of the initiative to fruition by sharing their experiences, lessons learned, and plans for next steps in creating a public, residential STEM school for highly able students. They hope that the opportunity to tell part of the story about the evolution of the Virginia STEAM Academy will be instructive for educators who aspire to pursue a similar vision in creating a STEM educational program.

With gratitude, I appreciate each invited author's expert contributions to this special text and thank the reader for her commitment in striving to continually provide better learning experiences for talented learners.

—Dr. Bronwyn MacFarlane, Editor

Reference

Gordon, E. (2009). The global talent crisis. *The Futurist,* September-October. Retrieved from http://www.wfs.org/node/877

PART I

Foundational Theories,
Research, and Best Practices
in STEM Education

CHAPTER 1

State Residential STEM Schools
A Case Study

Julia L. Roberts, Ed.D.

Schools with a focus on mathematics and science come in many forms, one of which is a residential school. Residential schools provide one of the educational services on the continuum planned to meet the unique needs of gifted adolescents (Cross & Miller, 2007; Rollins & Cross, 2014). Fifteen states have a state residential school that highlights science, technology, engineering, and mathematics (STEM). These schools provide opportunities for students to study at advanced levels and to do so with academic peers. Such schools allow students to accelerate their learning, especially in STEM subjects. Students at specialized state schools come from all corners of a state.

Three major reasons prompt establishing state residential STEM schools: One is educational, one relates to economic development, and the other is to keep exceptionally capable students in the state. The first goal is to offer opportunities for exceptional students to study at advanced levels. Many schools do not have the resources to develop the outstanding talents of students with interest in STEM careers. The second goal is to build a leadership cadre with expertise in STEM to promote economic development. The National Science Board (2010) put forth that "the long-term prosperity of our Nation will increasingly rely on talented and motivated individuals who will comprise the vanguard of scientific

 DOI: 10.4324/9781003238218-2

and technological innovation" (p. v). The third goal is to stem the "brain drain." State and local leaders are concerned about the number of students leaving the state with the fear that they will not return. These three reasons have prompted several states to establish state residential high schools with a focus on STEM.

Information About State Residential STEM Schools

Clarion Call For STEM Education

National reports have encouraged states to establish state residential schools.

Rising Above the Gathering Storm (National Academy of Sciences, 2007) recommended the expansion of two approaches to improving K–12 science and mathematics education that are already being implemented. One of the recommended approaches was establishing statewide specialty high schools: "Statewide specialty high schools can foster leaders in science, technology, and mathematics" (p. 9). Creating STEM-focused schools is one of the seven recommendations of the President's Council of Advisors on Science and Technology's (2010) *Prepare and Inspire: K–12 Education in Science, Technology, Engineering, and Math (STEM) for America's Future*. The report highlighted that "STEM-focused schools represent a unique National resource, both through their direct impact on students and as laboratories for experimenting with innovative approaches" (p. xii).

Brief History of State Residential STEM Schools

The first state residential STEM school was the North Carolina School of Science and Mathematics. It started in 1980 as an initiative of Governor James Hunt. The Illinois Mathematics and Science Academy opened in 1986, with the Texas Academy of Mathematics and Science getting its start in 1988. Since the 1980s, 12 other states have opened state residential STEM schools for a total of 15 states. Kentucky is the only state to have two such schools, with the second one planning to open in 2015. Table 1.1 details the schools.

TABLE 1.1

State Residential STEM Schools

School	Opening Year	Location	Website
Alabama School of Mathematics and Science	1991	Mobile, AL	http://www.asms.net
Arkansas School for Mathematics, Sciences, and the Arts	1993	Hot Springs, AR	http://www.asmsa.org
Craft Academy for Excellence in Science and Mathematics	2015	Morehead State University, Morehead, KY	http://www.moreheadstate.edu/craft-academy/
Gatton Academy of Mathematics and Science in Kentucky	2007	Western Kentucky University, Bowling Green, KY	http://www.wku.edu/academy
Georgia Academy of Aviation, Mathematics, Engineering, and Science	1997	Middle Georgia State College, Cochran, GA	http://www.mga.edu/games/
Illinois Mathematics and Science Academy	1986	Aurora, IL	http://www.imsa.edu
Indiana Academy for Science, Mathematics, and Humanities	1990	Ball State University, Muncie, IN	http://www.bsu.edu/academy
Kansas Academy of Mathematics and Science	2009	Fort Hays State University, Hays, KS	http://www.fhsu.edu/kams
Louisiana School for Math, Science, and the Arts	1983	Northwestern State University, Natchitoches, LA	http://www.lsmsa.edu
Maine School of Science and Mathematics	1995	Limestone, ME	http://www.mssm.org
Mississippi School for Mathematics and Science	1987	Mississippi University for Women, Columbus, MS	http://www.themsms.org
Missouri Academy of Science, Mathematics, and Computing	2000	Northwest Missouri State University, Maryville, MO	http://www.nwmissouri.edu/masmc/
North Carolina School of Science and Mathematics	1980	Durham, NC	http://www.ncssm.edu
Oklahoma School of Science and Mathematics	1990	Oklahoma City, OK	http://www.ossm.edu
South Carolina Governor's School for Science and Mathematics	1988	Hartsville, SC	http://www.scgssm.org
Texas Academy of Mathematics and Science	1988	University of North Texas, Denton, TX	http://rams.unt.edu

Creating STEM-focused schools is one of the seven recommendations of the President's Council of Advisors on Science and Technology's (2010) *Prepare and Inspire: K–12 Education in Science, Technology, Engineering, and Math (STEM) for America's Future*. The report highlighted that "STEM-focused schools represent a unique National resource, both through their direct impact on students and as laboratories for experimenting with innovative approaches" (p. xii).

What these schools have in common are the following: They are residential and open to students throughout the state, they receive support from the state budget, and students are selected based on their ability and interest in careers in science and mathematics. All of the schools include juniors and seniors, with a few also having sophomores in their student populations. A few of them also include the arts or humanities in their focus. Examples are the Arkansas School of Mathematics, Sciences and the Arts and the Indiana Schools of Science, Mathematics and Humanities.

Types of State STEM Schools

Two types of state residential STEM schools are represented in the 16 schools. The original model was the school with its own campus. The second model is the school located on a university campus. The free-standing schools have their own faculties and campuses. These schools have their own buildings and provide all services required by residential students. For example, they have their own laboratories, dining halls, and recreational facilities. Most of the state residential STEM schools represent this model.

The second model is characterized with a residence hall for the students on a university campus. Students take university courses with traditional college students and enjoy the services that are available on the campus. This model provides opportunities that are available on a university campus, such as visiting lecturers, musical events, recreational facilities, and expertise for research mentorships for students. This model also has the advantage of a lower budget.

Research on State Residential STEM Schools

Almarode, Subotnik, Crowe, Tai, Lee, and Nowlin (2014) stated, "Specialized science high schools offer an environment, both academic and social, in which interested students can explore the scientific world with both support and challenge" (p. 309). Rollins and Cross (2014) found "no evidence to support the notion that the residential school experience was harmful to student psychological development" (p. 337).

Wai, Lubinski, Benbow, and Steiger's (2010) longitudinal study showed that participating in numerous advanced precollegiate learning opportunities was linked to later accomplishments in STEM. The study by Almarode et al. (2014) found "49.8% of the selective STEM school graduates completed an undergraduate STEM degree" (p. 321) compared with 22.6% of all U.S. students entering college who complete a STEM undergraduate degree (National Science Board, 2012). Almarode et al. (2014) found that "a student's feelings of intellectual capacity in high school and the stability of interest in STEM related areas are strongly and positively associated with their persistence and earning an undergraduate degree in STEM" (p. 327).

> The study by Almarode et al. (2014) found "49.8% of the selective STEM school graduates completed an undergraduate STEM degree" (p. 321) compared with 22.6% of all U.S. students entering college who complete a STEM undergraduate degree (National Science Board, 2012).

A Case Study: The Gatton Academy of Mathematics and Science in Kentucky

The Carol Martin Gatton Academy of Mathematics and Science in Kentucky is an example of a state residential school that is located on the campus of Western Kentucky University (WKU). Gatton Academy students live in a specially designed residence hall and learn in university classes. There are many similarities with other state residential STEM schools, yet there are differences as well.

Getting Started

Dr. Julia Link Roberts, Director of The Center for Gifted Studies at WKU, and Dr. Charles McGruder, professor of physics and astronomy at WKU, submitted a proposal to the Kentucky Council of Higher Education in 1998. This proposal was to study the concept of having a state residential STEM school in Kentucky. Almost 10 years passed between the submission of this proposal and the opening of the doors to the first students at The Gatton Academy.

Years of advocacy and planning took place between the 1998 proposal and the state's eventual adoption of the idea. Roberts (2010) provided details of the advocacy and planning that led to a state residential STEM school being included in the state budget. Important decision makers to be informed about this opportunity included the candidates for governor, legislators, superintendents of school districts, and others in positions of influence.

In 2005, the state budget included funds to renovate a residence hall on the campus of WKU. In August 2007, The Gatton Academy opened, and Florence Schneider Hall became the home of The Gatton Academy and The Center for Gifted Studies. In 2014, funding was approved at the state level that will allow reaching the initial vision of 200 total students. An expansion of Schneider Hall will add student rooms and a community space large enough for all students and staff to gather for seminars and other total group meetings.

Sending Schools

The Gatton Academy provides a way for educators across the Commonwealth of Kentucky to extend advanced learning opportunities for high school students. Students who come to The Gatton Academy remain dually enrolled with their sending high schools. The state's per pupil funding continues to go to the sending school. All students at The Gatton Academy take the state assessment, and their scores are included with scores at the home high schools. In addition, educators at the sending schools are encouraged to celebrate honors and awards that the students receive, honoring the years spent learning during the previous years. These benefits were put in place to recognize and honor the local school districts and to assuage educators' concerns about their student(s) going to The Gatton Academy.

Dual enrollment is also a benefit to Gatton Academy students. It keeps the students connected to their home area. It allows the student to participate in commencement and other special occasions with students with whom they grew up.

The Application Process

Students around the state apply for admission to The Gatton Academy in a process similar to applying to a selective college. The online application is available from the beginning of the school year and is due on February 1. Students apply as sophomores, and they must be Kentucky residents. Included in the application submission are the student's transcripts from the freshman and sophomore years, ACT or SAT scores, letters of recommendation, and essays. Approximately 100 applicants are invited to come for a day of interviews as the final step in the process. An important part of the application process is the applicant's interest in pursuing a career in one or more STEM disciplines. Applicants share their interests in STEM as well as potential career goals through essays and interviews. Approximately 60 rising juniors receive invitations to join the upcoming class at The Gatton Academy. With the planned expansion, the number in each class will increase to 100 annually.

Student Benefits

The Gatton Academy "allows [students] to engage in learning at levels at which most of their age-mates are not yet ready to learn" (Roberts, 2013, p. 199). Students at The Gatton Academy graduate from high school with a minimum of 60 undergraduate hours. The state budget provides tuition, room, and board for students. Tuition allows students to take up to 19 credits of college coursework each semester during their junior and senior years of high school. Housing is provided with other Gatton Academy students, and a meal card allows students to eat at various locations on campus. Students are issued a laptop that includes software needed for classwork. Other benefits include the living-learning community, curriculum that offers advanced learning while still in high school, research opportunities, global experiences, and extracurricular opportunities.

A community of learners. Gatton Academy students live together in Florence Schneider Hall, a 1928 building that was renovated specifically to be the home of The Gatton Academy. The building is located in the heart of the campus of WKU. The students live "Shaker style" with the girls on one side of the building and boys on the other. There are the same number of rooms for males and females, so the gender balance is always in place.

Community spaces allow students to gather for studying and socializing. There are community spaces on three of the four floors in between the wings. In addition, space on each wing has a table, chairs, and sofas to provide another option for studying and gathering together. The fourth floor has a space large enough for all of The Gatton Academy students to hold seminars or for other occasions for all students. The expansion of the building will include a space for 200 students to get together for seminars, guest speakers, or other purposes.

> Educators at the sending schools are encouraged to celebrate honors and awards that the students receive, honoring the years spent learning during the previous years. These benefits were put in place to recognize and honor the local school districts . . . . Dual enrollment is also a benefit to Gatton Academy students. It keeps the students connected to their home area. It allows the student to participate in commencement and other special occasions with students with whom they grew up.

Students value opportunities to learn with others who are equally interested in learning. Most have never needed to ask for help; however, in this learning-living environment, they have ready access to others when they have questions or need assistance. They appreciate opportunities to learn with ideamates, others who are passionate about one or more areas of science, technology, engineering, and/or mathematics.

Special arrangements for safety are made for these students who are younger than most university students. A curfew is in place, and 24-hour surveillance is provided. A person is at the desk the whole time students are in residence. A residential counselor, who is a college graduate, lives on each wing of the building with Gatton students. Also, students scan their cards when they enter and exit the building.

Gatton Academy students do not have cars. The exception to the no-driving regulation is the few students who have permission to use a vehicle for transportation home on closed weekends. Those cars are parked at a remote parking lot and only accessed for driving to and from their homes. Approximately once a month, the building closes and students go home for the weekend with their families. Some of those closed weekends coincide with scheduled breaks for holidays.

Curriculum. The curriculum at the state residential STEM schools addresses the requirements for high school students and extends opportunities to learn at advanced levels. High school credits that are not yet met will be addressed with college courses in those curricular areas. All classes at The Gatton Academy are university classes taught by professors. Almost all classes are taken with regular college students. The exception would be a mathematics class in which there are so many Gatton Academy students that they fill a class, or the Computational Problem Solving class that was designed for and is a requirement for students at The Gatton Academy.

Gatton students begin mathematics study at the level at which they are prepared to start. Opportunities at sending high schools determine the starting class for most students. The math requirement for applying is the completion of Algebra II and Geometry. Some incoming students have completed more advanced math courses, so they may start with Calculus I or II. Appropriate placement of students in the math sequence is key to success in the study of mathematics.

All students take one course each in biology, chemistry, computer science, and physics during their 2 years at The Gatton Academy. In addition, they are required to select one of those content areas and take a second course in that sequence. They also must take three to four STEM electives from agriculture, architecture, astronomy, biology, chemistry, computer science, engineering, geography, geology, health sciences, manufacturing, mathematics, meteorology, physics, and/or psychological science.

A program option for Gatton Academy students is the STEM+Critical Language program, where students have the opportunity to study and master a critical language. The two critical languages from which students choose are Mandarin Chinese and Arabic. The Class of 2015 had eight students taking intensive language courses in Mandarin Chinese or Arabic. Students achieve a higher level of mastery of the language each semester. Students studying Chinese have the opportunity to matriculate into the Chinese Flagship Program at WKU after their first year of studying the language. Students have opportunities for language immersion in summer travel experiences. The STEM+Critical Language track is considered an equivalent to engaging in independent, mentored research.

Research opportunities. One of the exceptional opportunities for Gatton Academy students is the opportunity to engage in research from the time they enter the residential program. Although research is not a requirement for Gatton Academy students, 95% of the Class of 2014 participated in research. Gatton students engage in research with university faculty as mentors. Information about undergraduate research opportunities is shared with students at the beginning of their Gatton Academy experience. During Adventure Week (orientation for new students), WKU professors come to Florence Schneider Hall to highlight and share their research interests. Students may follow up by approaching the professors to serve as their mentors. Areas of potential research include agriculture, architecture, astronomy, biotechnology, biodiversity, bio-informatics, chemistry, computer science, civil engineering, electrical engineering, mechanical engineering, geography, geology, manufacturing, mathematics, physics, and psychology. Research is presented at university, state, and national conferences. A few Gatton students have published their research in professional journals.

> A program option for Gatton Academy students is the STEM+Critical Language program, where students have the opportunity to study and master a critical language. The two critical languages from which students choose are Mandarin Chinese and Arabic.

Summer research opportunities are supported by a gift from Mr. Carol Martin Gatton. This support is available for conducting research with a Western Kentucky University professor as mentor or elsewhere at a research facility. These internships are provided during the summer between the students' junior and senior years at the Academy. Examples of summer research included using

One of the exceptional opportunities for Gatton Academy students is the opportunity to engage in research from the time they enter the residential program. Although research is not a requirement for Gatton Academy students, 95% of the Class of 2014 participated in research. Gatton students engage in research with university faculty as mentors.

Google Glass to develop an app for WKU, lung cancer metabolism, and using the micro-EDM to study aerospace materials.

Quality research experiences are important for premier scholarship competitions. Submitting research results was the first step in the 2014 Siemens Competition of Math, Science & Technology. Gatton Academy students were semifinalists in the Siemens Competition with research topics that included studying micro-sized nuclear power sources, plasma physics for magnetic fusion energy, and calculation of stationary scattering states in 1D problems. Other research competitions also are open to high school students and undergraduate students, and Gatton Academy students usually qualify for both categories of competitions.

Global experiences. Students have three primary travel experiences in which they may participate. During the winter term, some Gatton Academy students participate in research in the rain forest in Costa Rica or travel to Italy or Greece (alternate years). Another study abroad experience occurs in England in the summer. Students spend 3 weeks at Harlaxton College in Grantham, England, where they study English literature and travel to sites of literary significance.

Gatton Academy students have spent their summers doing research in countries around the world. Others have immersed themselves in the study of a language in another country. Becoming a global citizen is an expectation of Gatton Academy graduates. As students are preparing for leadership roles in STEM disciplines, it is important to have a global perspective.

Quality research experiences are important for premier scholarship competitions. Submitting research results was the first step in the 2014 Siemens Competition of Math, Science & Technology. Gatton Academy students were semifinalists in the Siemens Competition with research topics that included studying micro-sized nuclear power sources, plasma physics for magnetic fusion energy, and calculation of stationary scattering states in 1D problems.

Extracurricular opportunities. Students engage in a variety of extracurricular activities, both at the high school and college level. They have activities within The Gatton Academy, and they also participate in extracurricular activities at WKU. They host a mid-winter dance as well as a prom. Social activities are available on weekends throughout the year. Students may audition for the orchestra, band, and choral groups at WKU. Opportunities for developing leadership and participating in interest groups or clubs are readily available.

Staff at The Gatton Academy

Staffing for The Gatton Academy is planned to offer needed support for students and to manage the ongoing operations of the state residential STEM school. The staff specializes in counseling, academic support, as well as student life. Each has specific responsibilities, and all work together to encourage and support the students.

Gatton Academy students have spent their summers doing research in countries around the world. Others have immersed themselves in the study of a language in another country. Becoming a global citizen is an expectation of Gatton Academy graduates. As students are preparing for leadership roles in STEM disciplines, it is important to have a global perspective.

Statewide Reach

The Gatton Academy is a statewide school. Consequently, it is important that students represent the entire state. In the 8-year history of The Gatton Academy, 113 of Kentucky's 120 counties have had one or more students as a student.

Getting the word out about the opportunity to apply for The Gatton Academy is an ongoing process. Sessions are held throughout the state at professional meetings and at sessions offered within regional communities. Mailings also provide information about the opportunities available at the state residential STEM high school.

Programming for Younger Students

"While the majority of the residential STEM schools have developed outreach programs throughout their histories, The Gatton Academy of Mathematics and Science in Kentucky represents a novel inversion of the trend" (Roberts & Alderdice, 2015). The Center for Gifted Studies at Western Kentucky University had more than a two-decade history of offering summer and Saturday programs to students in grades 1–10 when The Gatton Academy opened in 2007. The Center for Gifted Studies offers residential and nonresidential programming, including VAMPY, a 3-week program for seventh through tenth graders, and SCATS, a 2-week camp for sixth through eighth graders. The Gatton Academy and The Center for Gifted Studies share space at Florence Schneider Hall. The missions of both are complementary, and staff members provide support for each other.

Recognition

Newsweek and *The Daily Beast* recognized The Gatton Academy as the number one public high school in the United States in 2012 and 2013. *The Daily Beast* continued that top recognition in 2014. In the 2012 Intel Schools of Distinction Recognition, The Gatton Academy was named one of the three outstanding high school programs in the country. Such recognition was never a goal, yet the recognition of the programming provided by The Gatton Academy has helped spread the word about this educational opportunity.

Newsweek and *The Daily Beast* recognized The Gatton Academy as the number one public high school in the United States in 2012 and 2013. *The Daily Beast* continued that top recognition in 2014. In the 2012 Intel Schools of Distinction Recognition, The Gatton Academy was named one of the three outstanding high school programs in the country.

Expansion of The Gatton Academy

The original goal for The Gatton Academy was to have the capacity for 200 students—100 juniors and 100 seniors. When the original funding and bonding were available, the amount allowed for keeping the footprint of Florence Schneider Hall and renovating to accommodate 120 students. The number of Gatton Academy students will go up by 40 in the fall of 2016 and then to the full capacity of 200 students for the 2017–2018 academic year. The expansion was made possible by a generous gift and increased funding in the state budget in order to support the increased number.

Concluding Remarks

Fifteen states have implemented residential schools with a focus on STEM. These schools span a 35-year history. Such schools address the needs of students who are ready to learn at advanced levels that are not available at their sending high schools and who benefit from having proximity to academic peers or idea-mates.

Students who thrive at a state residential high school are eager to learn at advanced levels, enjoy the living-learning community, and are ready to take charge of both aspects of their lives. With the supports that are in place at The Gatton Academy, these young people achieve at high levels and engage in extracurricular activities, research, and global experiences. States benefit as these young people are interested in pursuing careers in science, technology, engineering, or mathematics.

> Fifteen states have implemented residential schools with a focus on STEM. These schools span a 35-year history. Such schools address the needs of students who are ready to learn at advanced levels that are not available at their sending high schools and who benefit from having proximity to academic peers or idea-mates.

Discussion Questions

1. What are elements of programming at The Gatton Academy that are appealing to students of advanced abilities and high interest in STEM content areas?
2. Which of these elements can be implemented in a nonresidential setting, and why would you want to do that?
3. How could the advocacy efforts that resulted in The Gatton Academy be generalized to something you want to implement in your school or school district?
4. How can STEM programming at The Gatton Academy or at a state residential school near you impact STEM programming in your school or school district?

References

Almarode, J. T., Subotnik, R. F., Crowe, E., Tai, R. H., Lee, G. M., & Nowlin, F. (2014). Specialized high schools and talent search programs: Incubators for adolescents with high ability in STEM disciplines. *Journal of Advanced Academics, 25*(3), 307–331.

Cross, T. L, & Miller, K. (2007). An overview of three models of publicly funded residential academies for gifted adolescents. In J. L. VanTassel-Baska (Ed.), *Serving gifted learners beyond the traditional classroom: A guide to alternative programs and services* (pp. 81–104). Waco, TX: Prufrock Press.

National Academy of Sciences. (2007). *Rising above the gathering storm: Energizing and employing America for a brighter economic future.* Washington, DC: The National Academies Press.

National Science Board. (2010). *Preparing the next generation of STEM innovators: Identifying and developing our Nation's human capital.* Arlington, VA: National Science Foundation.

National Science Board. (2012). *Science and engineering indicators 2012* (NSB 12-01). Arlington, VA: National Science Foundation.

President's Council of Advisors on Science and Technology. (2010). *Prepare and inspire: K–12 Education in science, technology, engineering, and math (STEM) for America's future.* Washington, DC: Author.

Roberts, J. L. (2010, Jan.). Lessons learned: A case study of advocating for a specialized school of mathematics and science, *Roeper Review, 32*(1), 42–47.

Roberts, J. L. (2013). The Gatton Academy: A case study of a state residential high school with a focus on mathematics and science. *Gifted Child Today, 36*(3), 193–200.

Roberts, J. L., & Alderdice, C. T. (2015). STEM-specialized schools. In S. G. Assouline, N. Colangelo, J. VanTassel-Baska, & A. E. Lupkowski-Shoplik (Eds.), *A nation empowered: Evidence trumps the excuses that hold back America's brightest students* (Volume II, 137–151). Iowa City: University of Iowa, The Belin-Blank Center for Gifted and Talented Education.

Rollins, M. R., & Cross, T. L. (December, 2014). Assessing the psychological changes of gifted students attending a residential high school with an outcome measurement. *Journal for the Education of the Gifted, 37*(5), 337–354.

Wai, J., Lubinski, D., Benbow, C. P., & Steigner, J. H. (2010). Accomplishment in science, technology, engineering, and mathematics (STEM) and its relation to STEM educational dose: A 25-year longitudinal study. *Journal of Educational Psychology, 102*(4), 860–871.

CHAPTER 2

S is for Science Education at the Elementary Level

Debbie Dailey, Ed.D.

A recent report indicated that 20% of all U.S. jobs required a significant background in at least one area of STEM (science, technology, engineering, mathematics; Rothwell, 2013). In 2011 alone, 26 million jobs in the U.S. were dependent upon workers competent in a STEM discipline. Unfortunately, our education system appears to be lagging in our efforts to prepare students for STEM-related careers. In a 2012 *Vital Signs* report, Change the Equation found that STEM job postings outnumbered unemployed people almost two to one. This gap was even wider in healthcare jobs with over three job postings for every unemployed person.

With the increasing demand for skilled STEM workers in our nation's workforce, our advanced learners, in particular, need opportunities to actively engage in the practices of science. To advance the development of STEM innovators, the National Science Board (NSB, 2010) recommended that K–12 students have investigative, real-world experiences in STEM learning and that these opportunities begin in the early grades. To facilitate this process, NSB suggested elementary teachers engage in professional development programs that support investigative classrooms and the identification of promising STEM learners.

 DOI: 10.4324/9781003238218-3

Early on, Brandwein (1995) understood the necessity of beginning science talent development in the early years. He proposed that elementary students be afforded investigative opportunities to encourage curiosity and engagement so that their interest in science would continue as they progressed in grade levels. Regrettably, many students are in middle school before they have opportunities to engage in the practices of science (Griffith & Scharmann, 2009) and by then, many learners have lost their initial interest (Maltese & Tai, 2011). In a recent study, Maltese and Tai reemphasized the necessity of early experiences in science. In interviews with 85 scientists and science graduate students, Maltese and Tai found that 65% of those interviewed developed their interest in science before middle school.

Research Supporting Engaging Science Practices

Curriculum

To improve the opportunities for advanced learners in science, teachers need access to a strong curriculum (Anderson, 2007; Brandwein, 1995; VanTassel-Baska, 1998). Subotnik, Olszewski-Kubilius, and Worrell (2011) believed the strength of the curriculum influences a student's interest in specific talent domains, such as science. To increase student interest and motivation to do science, Robinson, Shore, and Enerson (2007) suggested the curriculum address the content standards, focus on concepts, and employ an investigatory approach. VanTassel-Baska (1998) also recommended that science curricula emphasize inquiry, conceptual learning, higher level thinking, technology, and science process skills. VanTassel-Baska further stated the science curriculum should employ problem-based learning using real-world problems. Additionally, Robbins (2011) suggested the curriculum provide opportunities for students to (a) apply scientific reasoning to understand the world, (b) participate in reflection and collaboration with regard to science activities, (c) utilize quantitative methods to analyze scientific evidence, and (d) encounter the real work of scientists (e.g., through field-trip experiences).

Typically, science textbooks cover as many topics as possible, focusing on the width of science concepts as opposed to the depth of a concept. The National Research Council (NRC) reported that in comparison with nations that performed well on international science tests, the curricula in the United States did not target science concepts or concept linkage across grade levels but instead focused on a broad number of topics with little in-depth investigation (Duschl, Schweingruber, & Shouse, 2007). Due to this finding, NRC recommended that science instruction utilize a spiral curriculum that successively builds upon concepts across grade levels. In other words, NRC suggested students fully explore a few scientific concepts throughout the year instead of superficially covering many concepts as found in a typical science textbook.

Robbins (2011) suggested the curriculum provide opportunities for students to (a) apply scientific reasoning to understand the world, (b) participate in reflection and collaboration with regard to science activities, (c) utilize quantitative methods to analyze scientific evidence, and (d) encounter the real work of scientists (e.g., through field-trip experiences).

To build links across the curriculum, Metz (2008) recommended the curriculum include *big ideas* or *overarching concepts*, such as change, systems, scale, and evolution. Furthermore, researchers suggested overarching concepts be investigated at length and increased in complexity across succeeding grade levels (Duschl et al., 2006; Metz, 2008; Van Tassel-Baska, 1998, 2011). In support of increasing the depth of the science curriculum, Cotabish, Dailey, Robinson, and Hughes (2013) and Roth et al. (2011) reported increased student achievement when students made content connections utilizing overarching concepts.

When considering all students, but particularly culturally diverse and advanced learners, Lee and Buxton (2010) and Kim et al. (2012) suggested the science curriculum utilize inquiry and be relevant to students. For example, Project Clarion, an experimental study on the effects of an inquiry-based curriculum on elementary students' science learning, demonstrated increased science achievement for all students, particularly those from low socio-economic backgrounds (Kim et al., 2012). These promising results were also demonstrated among advanced learners. Robinson, Dailey, Hughes, and Cotabish (2014) and Feng, VanTassel-Baska, Quek, Bai, and Oneill (2005) found positive results among advanced learners after utilizing a curriculum involving inquiry and problem-based learning. As the results from the above studies indicated, curriculum based on understanding concepts, real-world problem solving, and inquiry-based learning benefited all learners but particularly high-ability learners.

Instructional Strategies

Inquiry-based science. The National Research Council (NRC, 1996) described scientific inquiry as the processes in which scientists' explain the natural world based upon experimental evidence. The NRC (2012) recommended that students engage in scientific inquiry to deepen their comprehension of the practices of science and to strengthen their conceptual understanding of science knowledge. They identified the following scientific practices essential to K–12 science: (a) posing authentic questions and defining problems; (b) creating and using models for explanations; (c) designing and conducting investigations; (d) analyzing and interpreting data; (e) using mathematics in problem solving; (f) using evidence to construct explanations or design solutions; (g) supporting arguments with evidence; and (h) interpreting, evaluating, and communicating results and conclusions.

Despite the hesitancy of elementary schools to adopt an inquiry-based curriculum due to testing demands (Anderson, 2007; Keil, Haney, & Zoffel, 2009), many studies supported the use of inquiry-based instruction for its impact on student achievement (Lynch, Kuipers, Pyke, & Szesze, 2005; Marx et al., 2004; Raghavan, Cohen-Regev, & Strobel, 2001; Young & Lee, 2005). In particular, a meta-analysis of 138 studies on inquiry-based instruction by Minner et al. (2010) identified overall positive results on students' science content knowledge, retention, and concept knowledge. Moreover, in an experimental study conducted by Ying-Tien and Chin-Chung (2005), researchers found an increase in student ability to describe, infer, and explain science concepts when students received inquiry-based instruction as compared to students receiving traditional instruction.

Problem-based learning (PBL). Inquiry learning is highlighted in many different instructional strategies including problem-based learning (Chin & Chia, 2004; Furtado, 2010; Hmelo-Silver, Duncan, & Chinn, 2006). According to Drake and Long (2009), problem-based learning experiences allow students to gain content understanding through the process of inquiry. Problem-based learning appears to be a perfect fit for science (Akinoglu & Tandogan, 2007) and high-ability learners (Gallagher, 2005) because of the focus on active learning, experimental evidence, collaboration among stakeholders, and real-world issues. Inel and Balim (2010) described a typical PBL experience being framed around an event about daily life in which students work collaboratively as scientists to define, research, and form solutions to a problem. In addition, students construct experiments to test the effectiveness of their proposed solutions. Throughout the PBL experience, students impersonate various stakeholders to present varied sides of the problem and use argumentation, supported with experimental evidence, to further their position (Belland, Glazewski, & Richardson, 2011; VanTassel-Baska, 2011).

Researchers documented the positive effects of PBL on student achievement and enthusiasm for learning. Akinoglu and Tandogan (2007) reported statistically significant gains on measures of science achievement and student attitude toward science among students who received science instruction through PBL when compared with students in a traditional science class. Drake and Long (2009) found grade 4 students in PBL classrooms demonstrated increased content knowledge, student engagement, and problem-solving abilities compared to students in a traditional classroom. Keil et al. (2009) also reported increased scores on measures of student achievement and science process skills. A noteworthy finding included increased student achievement on state-mandated proficiency exams across various subject areas by students who participated in the PBL intervention. In summary, higher level thinking capabilities, such as problem solving, conceptual understanding, and process-oriented skills can be improved upon and facilitated through problem-based learning experiences (Akinoglu & Tandogan, 2007; Drake & Long, 2009; Gijbels, Dochy, Van den Bossche, & Segers, 2005; Sungar, Tekkaya, & Geban, 2006). Moreover, PBL facilitates the learning of soft skills, such as research, written and oral communication, and collaboration (Allen, Donham, & Bernhardt, 2011).

Barriers to Improving Science in the Elementary Grades

Even in our gifted programs, science is often seen as an afterthought and is not adequately supported. Callahan, Moon, and Oh (2014) found the majority of elementary gifted programs identified language arts as the most well-developed content area used in their program (47.2%) while a mere 10.5% of schools identified science and technology as the primary content area. The reasons for the lack of science in elementary classrooms and gifted programs are numerous but they include: (a) time constraints and scheduling conflicts, (b) insufficient resources, (c) inadequate teacher science knowledge and skills, and (d) poor teacher confidence.

Time constraints. Due to the demands of the No Child Left Behind Act and the stringent focus on test results, many school districts opted to reduce the amount of time spent on science in elementary classrooms (Griffith & Scharmann, 2008). The Center on Education Policy (CEP, 2006) reported that 68% of school districts reduced the time spent on science instruction to extend the time for language arts and mathematics. Sandholtz and Ringstaff (2011) found that 56% of the teachers indicated time as the greatest barrier to effective science instruction and once the testing season began, science was all but forgotten. Additionally, preparing to teach inquiry-based science required more preparation time than some teachers would sacrifice, especially when science was not part of Adequate Yearly Progress (AYP; Griffith & Scharmann, 2008). At grade levels where science is tested through state assessments, many teachers believed state science exams assessed broad, knowledge-based science facts as opposed to more in-depth application and analysis items utilized in inquiry science (Johnson, 2006). In this case, teachers worried they would not cover all required content if they took time to cultivate an investigative classroom using inquiry-based science (Johnson, 2006). In summary, the lack of time to prepare and implement inquiry-based science in elementary schools created a barrier to effective science instruction.

Resources

Many teachers in gifted programs and regular classrooms lack the resources, such as physical space, supplies, equipment, and appropriate curriculum, to effectively provide inquiry-based learning opportunities for students (Johnson, 2006). When science is taught, numerous teachers pay for materials out of their own pocket (Buczynski & Hansen, 2010), creating an additional barrier for teachers who do not have the personal financial resources to supplement their classrooms. Furthermore, the management of science equipment and the setting up of experimental activities create additional obstacles for teachers (Peers, Diezmann, & Watters, 2003).

Science Content Knowledge and Pedagogical Skills

The lack of content knowledge and pedagogical skills inhibits a teacher's ability to teach science, specifically inquiry-based science, to advanced learners (Coates, 2006). Teachers often express anxiety in teaching science due to their insufficient knowledge of the subject, thereby negatively influencing their instruction (Kallery, 2004). For example, teachers avoid inquiry-based instruction for fear their inadequate knowledge would lead students in the wrong direction. Coates found this barrier especially true when facilitating discussion and investigations with advanced learners.

Leading an inquiry-based science lesson presents another barrier, especially to teachers who prefer a more traditional or didactic classroom. Inquiry-based classrooms involve students in practices of science, such as experimental investigations, while utilizing minimal teacher-centered instruction (Harris & Rooks, 2010). According to Johnson (2006), teachers who learned science primarily through lecture and worksheets struggle with the types of instructional practices normally used in inquiry-based learning. Together, lack of content knowledge and pedagogical skills hampers a teacher's ability to facilitate inquiry-based learning in the classroom (Coates, 2006; Choi & Ramsey, 2009; Kallery, 2004; Peers et al., 2003), and furthermore, lessens a teacher's confidence in his or her ability to teach science (Goodnough & Nolan, 2008; Murphy, Neil, & Beggs, 2007).

Teacher Confidence

Confidence in teaching a subject affects how the subject will be taught. In particular, teachers who possess a low confidence in teaching science usually limit the amount of time, inquiry, and discussion spent on science (Jarvis & Pell, 2004; Murphy et al., 2007; Sinclair, Naizer, & Ledbetter, 2011). In a survey among 300 elementary teachers, responders indicated that their lack of confidence in teaching science had the largest impact on their science teaching (Murphy et al., 2007). In another study, Jarvis and Pell (2004) found teachers lacked confidence in teaching science as compared to other subjects, such as mathematics and English, especially when it came to teaching inquiry-based science. To increase teachers' confidence in teaching inquiry-based science, Murphy and colleagues (2007) suggested teachers participate in quality professional development that addresses their science knowledge and skills.

In summary, time constraints, resources, content knowledge, pedagogical skills, and confidence influence the frequency and quality of science instruction in both gifted and regular classrooms. To address these barriers, experts recommend school districts afford time in the day for science instruction, secure the necessary resources for an investigative classroom, and provide teachers with increased professional development opportunities that target content knowledge, pedagogical skills, and confidence in teaching science (Buczynski & Hansen, 2010; Duschl, et al., 2007; NSB, 2010; Robinson, Shore, & Enerson, 2007).

Practices to Best Serve
Gifted Learners in Science

Taking what the research says about effective science instruction for advanced learners, how can educators best develop the science talent of gifted learners? To facilitate interest and motivate gifted learners in science, educators should begin with relevant, meaningful experiences. Real-world problems, introduced through problem-based or project-based learning, allow students to place themselves in the center of the problem as they work to find possible solutions.

In problem-based learning, students are typically presented a real-world problem situated in some type of scenario. Scenarios should always be created with relevancy in mind. In other words, students' experiences, geographical location, and knowledge base should be considered. You would not want to introduce a scenario about struggles in the desert to students who live in an area of plentiful rainfall. Scenarios linked to community-based issues often work well in gifted classrooms and can lead to a service-learning project. For example, you might introduce a scenario and problem centered around an issue that you have on your school grounds . . . such as the flooding on the playground. Students would be presented with the problem and then act as investigators to find possible solutions to the problem. Depending on the feasibility of the solution, students could act to put the solution into effect.

> Real-world problems, introduced through problem-based or project-based learning, allow students to place themselves in the center of the problem as they work to find possible solutions.

Project-based learning is similar to problem-based learning but is more focused on creating a product, presentation, or performance (Adams, Cotabish, & Dailey, 2015). Again it is important to engage students in relevant, meaningful, and personally interesting projects. As in problem-based learning, the teacher acts as the facilitator as he or she guides students through the process. Project-based learning also utilizes scenarios usually in conjunction with a driving or essential question. A common scenario might be addressing the need for a walking bridge over a creek on your school grounds. The driving question, which is used to frame the project in the content, could be, "How can we design a safe and sturdy walking bridge for our school?" Depending on available resources, this project could be completed or students could be engaged in creating a prototype of the bridge. Once the prototype is completed, students might wish to present their plan to the school board to see if they can garner support for their project.

Regardless of whether teachers use problem or project-based learning, it is essential that they situate the learning in the content. If content is not involved then students are just doing problem-solving activities and projects. Considering the above problem-based scenario, teachers could tie in content standards that address water properties and the water cycle, erosion, and ecosystems, to name a few. Considering the above project-based scenario, teachers could tie in content standards that address balance and forces, engineering design practices, erosion, geography, economics, and more.

New Directions

With the release of the Next Generation Science Standards (NGSS), efforts are being made to improve science instruction and to provide students early opportunities to engage in the practices of science. NGSS frames science learning across three dimensions: Science and Engineering Practices, Crosscutting Concepts, and Disciplinary Core Ideas. The dimensions should not be taught in isolation, but rather be integrated together to explain a particular phenomenon. For example, as students learn about Newton's Third Law, students should connect the content to other cause and effect relationships as they experiment with action-reaction forces. This practice is not new to gifted educators. As early as 1992, VanTassel-Baska (1992, 1998) suggested science curriculum focus on three dimensions (advanced content, process/product, and overarching concepts) in her Integrated Curriculum Model. Through this type of model, conceptual understanding is deepened as students engage in deep exploration of content, supported with real-world investigations, that is linked to previous and future content by overarching concepts.

The NGSS can be used as a point of differentiation for gifted learners. The NGSS are informed by learning progressions that describe the pathway students will take to master a concept. These learning progressions increase in complexity as students advance in grade levels. For example, students in early grades will learn that organisms have external parts that they use to perform daily functions. As the grade levels advance and they continue their study of structure and function, students will eventually learn about specialized cells and subcellular particles that perform essential functions for life (NGSS Lead States, 2013). When students master a certain concept, teachers can use these learning progressions to accelerate the student beyond their grade level. The NGSS also include clarification statements and assessment boundaries to guide teachers in their lesson planning. Teachers of high-ability students can use these statements and boundaries to challenge learners beyond the expectations of typical learners (Adams, Cotabish, & Dailey, 2015; Adams, Cotabish, & Ricci, 2014).

Implications for Teacher Preparation of Gifted Learners

Many gifted teachers, particularly elementary teachers, did not specialize in science, nor do they have a strong science background. The curriculum can be great but without an effective teacher, science instruction will be marginal. The key to effectively engaging gifted learners in science involves improving the science teaching practices of their teachers. Critical elements of professional development should provide teachers with extended contact time, follow-up support, explicit instruction on teaching practices using classroom-specific curriculum, and should address barriers to science instruction (Cotabish, Dailey, Robinson, Hughes, 2013; Dailey & Robinson, 2013).

One-shot workshops of the past will not effectively address these needs. To enact real changes in the science classroom, stakeholders must be willing to provide support and opportunities for teachers to be trained and to provide materials and resources necessary to create ideal science-learning opportunities for students.

Discussion Questions

1. In many states, science is not a tested subject until middle school. What can teachers of gifted students do to encourage a greater focus on science in their school systems? Why should they be concerned with a lack of focus on science?
2. Considering the enormous demands on teachers, how can school districts and administrators encourage general and gifted elementary teachers to give up time in their summers to attend professional development institutes focused on science? What other options besides summer would be feasible and would address the critical elements of professional development?

References

Adams, A. Cotabish, A., & Dailey, D. (2015). *A teacher's guide to using the Next Generation Science Standards with gifted and advanced learners*. Waco, TX: Prufrock Press.

Adams. C., Cotabish, A., & Ricci, M. K. (2014). *Using the Next Generation Science Standards with advanced and gifted learners*. Waco, TX: Prufrock Press.

Akinoglu, O., & Tandogan, R. O. (2007). The effects of problem-based active learning in science education on students' academic achievement, attitude, and concept learning. *Eurasia Journal of Mathematics, Science and Technology Education, 3*(1), 71–81.

Allen, D. E., Donham, R. S., & Bernhardt, S. A. (2011). Problem-based learning. *New Directions for Teaching and Learning, 128,* 21–29.

Anderson, R. D. (2007). Inquiry as an organizing theme for science curricula. In S. K. Abell & N. G. Lederman (Eds.), *Handbook of research on science education* (pp. 807–830). Nahway, NJ: Lawrence Erlbaum Associates.

Belland, B. R., Glazewski, K. D., & Richardson, J. C. (2010). Problem-based learning and argumentation: Testing a scaffolding framework to support middle school students' creation of evidence-based arguments. *Instructional Science, 39,* 667–694.

Buczynski, S., & Hansen, C. B. (2010). Impact of professional development on teacher practice: Uncovering connections. *Teacher and Teacher Education, 26,* 599–607.

Brandwein, P. F. (1995). *Science talent in the young expressed within ecologies of achievement* (RBDM 9510). Storrs: University of Connecticut, The National Research Center on the Gifted and Talented.

Callahan, C., Moon, T., & Oh, S. (2014). *Status of elementary gifted programs.* Retrieved from http://nagc.org/uploadedFiles/Information_and_Resources/ELEM%20school%20GT%20Survey%20Report.pdf

Center on Education Policy. (2006, March). *From the capital to the classroom: Year 4 of the No Child Left Behind Act summary and recommendations.* Retrieved from http://www.cep-dc.org/displayDocument.cfm?DocumentID=301

Change the Equation. (2012). *Vital signs: Reports on the condition of STEM learning in the U.S.* Retrieved from http://changetheequation.org/sites/default/files/ CTEq_VitalSigns_Supply (2).pdf

Chin, C., & Chia, L. G. (2004). Problem-based learning: Using students' questions to drive knowledge construction. *Science Educator, 88*, 707–727.

Choi, S., & Ramsey, J. (2009). Constructing elementary teachers' beliefs, attitudes, and practical knowledge through an inquiry-based elementary science course. *School Science and Mathematics, 109*, 313–324.

Coates, D. (2006). 'Science is not my thing': Primary teachers' concerns about challenging gifted pupils. *Education, 34*, 49–64.

Cotabish, A., Dailey, D., Robinson, A., & Hughes, G. (2013). The effects of a STEM intervention on elementary students' science knowledge and skills. *School Science and Mathematics, 113*(5), 215–226.

Dailey, D., & Robinson, A. (2013). The effect of implementing a STEM professional development intervention on elementary teachers. Retrieved from ProQuest (UMI 3587609).

Drake, K. N., & Long, D. (2009). Rebecca's in the dark: A comparative study of problem-based learning and direct instruction/experimental learning in two 4th-grade classrooms. *Journal of Elementary Science Education, 21*(1), 1–16.

Duschl, R., Schweingruber, H. A., & Shouse, A. (2007). *Taking science to school: Learning and teaching science in grades K–8*. Washington, DC: The National Academies Press.

Feng, A. Z., VanTassel-Baska, J., Quek, C., Bai, W., & Oneill, B. (2005). A longitudinal assessment of gifted students' learning using the integrated curriculum model (ICM): Impacts and perceptions of the William and Mary language arts and science curriculum. *Roeper Review, 27*(2), 78–83.

Furtado, L. (2010). Kindergarten teachers' perceptions of an inquiry-based science teaching and learning professional development intervention. *New Horizons in Education, 58*(2), 104–120.

Gallagher, S. A. (2005). Adapting problem-based learning for gifted students. In F. A. Karnes & S. M. Bean (Eds.), *Methods and materials for teaching the gifted* (2nd ed., pp. 285–312). Waco, TX: Prufrock Press.

Gijbels, D., Dochy, F., Van den Bossche, P., & Segers, M. (2005). Effects of problem-based learning: A meta-analysis from the angle of assessment. *Review of Educational Research, 75*(1), 27–61.

Goodnough, K., & Nolan, B. (2008). Engaging elementary teachers' pedagogical content knowledge: Adopting problem-based learning in the context of science teaching and learning. *Canadian Journal of Science, Mathematics, and Technology Education, 8*, 197–216.

Griffith, G., & Scharmann, L. (2008). Initial impacts of No Child Left Behind on elementary science education. *Journal of Elementary Science Education, 20*(3), 35–48.

Harris, C. J., & Rooks, D. L. (2010). Managing inquiry-based science: Challenges in enacting complex science instruction in elementary and middle school classrooms. *Journal of Science Teacher Education, 21,* 227–240.

Hmelo-Silver, C. E., Duncan, R. G., & Chinn, C. A. (2006). Scaffolding and achievement in problem-based and inquiry learning: A response to Kirschner, Sweller, and Clark. *Educational Psychologist, 42*(2), 99–107.

Inel, D., & Balim, A. G. (2010). The effects of problem-based learning in science and technology teaching upon students' academic achievement and levels of structuring concepts. *Asia-Pacific Forum on Science Learning and Teaching, 11*(2), 1–23.

Jarvis, T., & Pell, A. (2004). Primary teachers' changing attitudes and cognition during a two-year science in-service program and their effect on pupils. *International Journal of Science Education 26,* 1787–1811.

Johnson, C. C. (2006). Effective professional development and change in practice: Barriers teachers encounter and implications for reform. *School Science and Mathematics, 106*(3), 1–26.

Kallery, M. (2004). Early years teachers' late concerns and perceived needs in science: An exploratory study. *European Journal of Teacher Education, 27,* 147–165.

Keil, C., Haney, J., & Zoffel, J. (2009). Improvements in student achievement and science process skills using environmental health science problem-based learning curricula. *Electronic Journal of Science Education, 13*(1), 1–18.

Kim, K. H, VanTassel-Baska, J., Bracken, B. A., Feng, A., Stambaugh, T., & Bland, L. (2012). Project Clarion: Three years of science instruction in Title 1 schools among K–Third grade students. *Research in Science Education. 42,* 813–829.

Lee, O., & Buxton, C. A. (2010). Diversity and equity in science education: Research, policy, and practice. New York, NY: Teachers College Press.

Lynch, S., Kuipers, J., Pyke, C., & Szesze, M. (2005). Examining the effects of a highly rated curriculum unit on diverse populations: Results from a planning grant. *Journal of Research in Science Teaching, 42,* 912–946.

Maltese, A. V., & Tai, R. H. (2010). Eyeballs in the fridge: Sources of early interest in science. *International Journal of Science Education, 32,* 669–685.

Marx, R., Blumenfeld, P. C., Krajcik, J., Fishman, B., Soloway, E., Geier, R., & Tal, R. T. (2004). Inquiry-based science in the middle grades: Assessment of learning in urban systemic reform. *Journal of Research in Science Teaching, 41,* 1063–1080.

Metz, K. E. (2008). Narrowing the gulf between the practices of science and the elementary school science classroom. *The Elementary School Journal, 109*(2), 138–161.

Minner, D. D., Levy, A. J., & Century, J. (2010). Inquiry-based science instruction: What is it and does it matter? Results from a research synthesis years 1984–2002. *Journal of Research in Science Teaching, 47,* 474–496.

Murphy, C., Neil, P., & Beggs, J. (2007). Primary science teacher confidence revisited: Ten years on. *Educational Research, 49,* 415–430.

National Research Council. (1996). *National science education standards.* Washington, DC: National Academy Press.

National Research Council. (2012). A framework for K–12 science education: Practices, crosscutting concepts, and core ideas. Committee on a Conceptual Framework for New K-12 Science Education Standards. Board on Science Education, Division of Behavioral and Social Science and Education. Washington, DC: The National Academies Press.

National Science Board. (2010). Preparing the next generation of STEM innovators: Identifying and developing our nation's human capital (NSB-10-33). Retrieved from http://www.nsf.gov/nsb/publications/2010/nsb1033

NGSS Lead States. (2013). *Next Generation Science Standards: For States, by States.* Washington, DC: The National Academies Press.

Peers, C. E., Diezmann, C. M., & Watters, J. J. (2003). Supports and concerns for teacher professional growth during the implementation of a science curriculum innovation. *Research in Science Education, 33,* 89–110.

Raghavan, R., Cohen-Regev, S., & Strobel, S. A. (2001). Student outcomes in a local systemic change project. *School Science and Mathematics, 101,* 417–426.

Robbins, J. I. (2011). Adapting science curricula for high-ability learners. In J. VanTassel-Baska & C. A. Little (Eds.), *Content-based curriculum for high-ability learners* (2nd ed., pp. 217–238). Waco, TX: Prufrock Press.

Robinson, A., Dailey, D., Hughes, G., & Cotabish, A. (2014). The effects of a science-focused STEM intervention on gifted elementary students' science knowledge and skills. *Journal of Advanced Academics, 25,* 159–161.

Robinson, A., Shore, B. M., & Enersen, D. L. (2007). *Best practices in gifted education: An evidence-based guide.* Waco, TX: Prufrock Press.

Roth, K. L., Garnier, H. E., Chen, C., Lemmens, M., Schwille, K., & Wickler, N. I. Z. (2011). Videobased lesson analysis: Effective science pd for teacher and student learning. *Journal of Research in Science Teaching, 48*(2), 117–148.

Rothwell, J. (2013). The hidden STEM economy. Washington, DC: Brookings Institute. Retrieved from http://www.brookings.edu/research/reports/2013/06/10-stem-economy-rothwell

Sandholtz, J. H., Ringstaff, C. (2011). Reversing the downward spiral of science instruction in K–2 classrooms. *Journal of Science Teacher Education, 22,* 513–533.

Sinclair, B. B., Naizer, G., & Ledbetter, C. (2011). Observed implementation of a science professional development program for K–8 classrooms. *Journal of Science Teacher Education, 22,* 579–594.

Subotnik, R. F., Olszewski-Kubilius, P., & Worrell, F. C. (2011). Rethinking giftedness and gifted education: A proposed direction forward based on psychological science. *Psychological Science in the Public Interest, 12,* 3–54.

Sungar, S., Tekkaya, C., & Geban, O. (2006). Improving achievement through problem-based learning. *Educational Research, 40*(4), 155–160.

VanTassel-Baska, J. (1992). *Planning effective curriculum for gifted learners.* Denver, CO: Love Publishing.

VanTassel-Baska, J. (1998). Planning science programs for high ability learners. ERIC Clearinghouse on Disabilities and Gifted Education. Retrieved from http://www.ericdigests.org/1999-3/science.htm

VanTassel-Baska, J. (2011). Implementing innovative curriculum and instructional practices in classrooms and schools: Using research-based models of effectiveness (pp. 437–465). In J. VanTassel-Baska & C. A. Little (Eds.), *Content-based curriculum for high-ability learners* (2nd ed., pp. 437–465). Waco, TX: Prufrock Press.

Ying-Tien, W., & Chin-Chung, T. (2005). Effects of constructivist-oriented instruction on elementary school students' cognitive structures. *Journal of Biological Education, 39,* 113–119.

Young, B. J., & Lee, S. K. (2005). The effects of a kit-based science curriculum and intensive science professional development on elementary student science achievement. *Journal of Science Education and Technology, 14,* 417–481.

CHAPTER 3

S is for Science Education at the Secondary Level

Steve V. Coxon, Ph.D.

A 9-month-old infant, riding in a shopping cart and dropping groceries on the floor to test gravity, has more in common with a professional scientist than most secondary students will experience in school. There is a wide chasm between secondary science instruction in most schools and the actual work of scientists. If middle and high school students are fortunate enough to attend a school with resources appropriated to science labs and materials—and many are not so lucky—they are usually made to follow step-by-step instructions. The best students in this model simply come to a result known by the teacher and long known to science. Although it is important to learn to follow instructions, develop lab skills, and have a base of scientific knowledge, these should be seen as the floor and not the ceiling. Many schools are training technicians and not preparing scientists. Scientists seek answers to yet unanswered questions and solutions to yet unsolved problems through rigorous methodology, including designing and conducting their own experiments.

Students at all levels, from preschool through high school and beyond, can design and conduct their own experiments to answer relevant questions about real-world problems and participate in this fundamentally Constructivist model of science instruction. This is in sharp contrast to the instruction in

 DOI: 10.4324/9781003238218-4

Students at all levels, from preschool through high school and beyond, can design and conduct their own experiments to answer relevant questions about real-world problems and participate in this fundamentally Constructivist model of science instruction.

the typical, traditional classroom in which a teacher and text are seen as the sole sources of information; work is individualized and simplistic in nature with close-ended, single-answer problems; and students are dominated by the teacher to restrict their movements, discussions, and habits. This model of education seems unlikely to sufficiently develop scientific thinking and habits or to inspire the most capable students to pursue science fields at the postsecondary level. Instead, secondary science classrooms should model real scientific environments: with multifaceted sources of information; teamwork; complex, real-world problems for which students develop and conduct experiments to solve facets; and teachers who serve as facilitators and guides.

Changing this paradigm in schools is of prime importance. According to the National Science Board (NSB; 2010), U.S. economic growth has been based on innovations in science since at least World War II. In particular, when the Soviets launched Sputnik through the American night sky in 1957, it spawned a flurry of interest and spending on science education, and the U.S. has arguably led the world in scientific productivity and innovation ever since. However, this may change. The National Academy of Sciences (NAS; 2007) warned that although the U.S. is performing well in scientific innovation internationally, other nations are catching up. The Business Roundtable (2006, as cited in NSB, 2010), argued that waiting for another Sputnik moment may result in a slow withering of U.S. scientific superiority. This is seen in the low number of young people choosing to enter scientific fields. Although the U.S. Bureau of Labor Statistics predicts STEM occupations will grow by 17% this decade, far too few high school graduates are pursuing STEM majors (Langdon, McKittrick, Beede, Khan, & Doms, 2011). Although 28% of college students begin as STEM majors, about half will either switch majors or drop out of school before graduating (Chen & Soldner, 2013). The NSB (2010) noted that it is important that high-ability students in particular enter science fields. Engaging, authentic science education in secondary schools could inspire and better prepare the most capable graduates to pursue science at the university level and beyond. In this chapter, key elements of secondary

Although 28% of college students begin as STEM majors, about half will either switch majors or drop out of school before graduating (Chen & Soldner, 2013). The NSB (2010) noted that it is important that high-ability students in particular enter science fields. Engaging, authentic science education in secondary schools could inspire and better prepare the most capable graduates to pursue science at the university level and beyond.

science curriculum are summarized, project- and problem-based learning are discussed as ideal methods for the secondary science classroom, and programming options including science-related advanced coursework, special schools, and extracurricular programs are discussed.

Science Curriculum for the 21st Century

VanTassel-Baska (1992) connected her Integrated Curriculum Model (ICM) to the special needs of gifted learners. Curriculum developed with the ICM includes advanced content, process-product, and concept dimensions. Each component of the ICM helps to meet the special needs of gifted learners. Advanced content, such as the Next Generation Science Standards (NGSS) and Advanced Placement (AP) classes, responds to gifted learners' educational precocity. The process-product dimension responds to gifted learners' intensity for learning through the use of rigorous processes, such as the scientific method to lead to real-world products (e.g., a symposium to present results). Conceptual understanding addresses gifted learners' complex thinking ability by addressing big ideas such as evolution, systems, and change.

Constructivism in the Secondary Science Classroom

The ICM partners well with Brooks and Brooks' (1993) 12-point framework of Constructivist classrooms, which is summarized here for the secondary science classroom:

1. Encourage autonomy and initiative. Science careers are some of the most open-ended careers, especially for university faculty, scientific innovators, and entrepreneurs. This requires intrinsic motivation be fostered in students, not authoritarianism and dictated step-by-step instructions.
2. Use manipulative, interactive, and physical materials. Hands-on work is not only important for children: Adolescents and adults learn better this way as well. It is vital that secondary students are offered expe-

riences to use and explore authentic lab materials and to pursue the life sciences in their fields of interest.

3. Use cognitive terminology. The language of science should be used and never dumbed down for students. Secondary students will be more college and career ready if they have opportunities to learn and use the language of science. This is especially important for students from poverty who are less likely to hear scientific terms outside of school.

4. Student needs drive lessons, strategies, and content. While science standards and the content of tests, such as AP tests, may drive the content of class, students can be given great autonomy in how they learn as described later with project- and problem-based learning. To the extent possible, students should be encouraged to explore content areas of interest.

5. Teachers discover students' understandings of concepts before sharing their own. This puts the onus of learning on the student and demands student thinking instead of passive absorption of teacher-directed material.

6. Encourage students to engage in discussion. Teachers should act as guides, asking questions to steer the conversation. For example, if a student has designed an experiment that is unfair or left out possible threats to the validity of their results, the teacher might ask directed questions to help the student realize the issue instead of merely telling them.

7. Encourage student inquiry. This corresponds strongly with number 4, and as students are given opportunities to explore the subject with great autonomy, questions will likely arise that are suitable for further exploration both through researching existing information with the Internet, the library, and expert knowledge, as well as through experimentation when the information is unknown.

8. Seek elaboration from students. Teachers should not accept superficial or low-level answers, but use questions to push and therefore deepen student thinking. Elaboration is a vital skill in the sciences, including in the creative process. Elaboration has decreased a greater amount than any other area of creativity in what has become known as the Creativity Crisis—the overall decline in creativity test scores over the past two decades (Kim & Coxon, 2013). Scientific innovation requires endurance to emerge with a final solution and product.

Real-world scientific problems are complex, and solutions to them must be multifaceted and well elaborated upon.

9.   Engage students in experiences that might engender contradictions. Real-world science problems are not clean and easy with textbook solutions, thus problems given in the classroom should be ill structured. That is, there should be too much information, some of which is not relevant. There should be many avenues to explore and many means by which to search for solutions.

10. Allow wait time after posing questions. As a teacher educator, I have found this to be very challenging for new teachers in particular. Although the contrasted traditional model of teaching lends itself toward quick recitation of memorized answers, the Constructivist classroom requires deep thinking, and this takes time. Successful scientists spend time considering difficult questions before proceeding; secondary classrooms should model this.

11. Provide time for students to construct meaning. As described above, understanding takes time. It also means that students should be given opportunities to try and fail without harming their grade. For example, if a student designs an experiment that fails to help answer the research question, opportunities should be provided to the extent possible to make adjustments and try again. Learning to persevere in the face of failure is a real-world skill: Scientists move on and learn from botched experiments, must revise and resubmit manuscripts to meet the demands of critical peers, and get rejected conference proposals. Attempted innovations are not always successful, patent requests may be denied, and entrepreneurial ventures often do not succeed. Successful scientists must learn from failure and move forward. Perseverance in the face of failure takes time and opportunities to develop; time too rarely offered in the traditional classroom.

12. Nurture students' natural curiosity. In my experience having taught elementary, middle, high school, undergraduate, and graduate university students, curiosity declines with age for most students. Humans are innately curious, but the traditional classroom seems likely to systematically reduce curiosity by encouraging the memorization of known information instead of encouraging discovery of the unknown. Curiosity is foundational to creativity and scientific innovation. Aristotle, Galileo, and Einstein were each intensely curious adults, as are most groundbreaking, living scientists. Secondary teachers may find it necessary to reinvigorate their students' curiosity depending

on the students' prior experiences. This can be done through modeling, including thinking aloud from the teacher, as well as through the methods described later in this chapter.

The 4 Cs

The Constructivist classroom described above and the ICM together align well with 21st-century learning, the **4 Cs: creativity, critical thinking, communication, and collaboration**. The Constructivist classroom provides the collaborative environment while the ICM provides the processes, content, and conceptual development that foster student growth in communication, creativity, and critical thinking. In the 21st-century secondary science classroom, **collaboration** means that students work side-by-side with other students in the classroom and with their teacher, as well as potentially with students far away and with experts on the topic at hand. **Communication** means speaking, listening, reading, and writing in many formats. For example, reading fiction and nonfiction require very different skills. Writing, in particular, is a vital skillset in which a high level of technical exactitude will distinguish those who are successful professionally from those who are not across disciplines. Being successfully **creative** requires the creation of something new or improved that has value (Kim & Coxon, 2013). In science, this could be an idea behind a technology product that will be later engineered. For example, smartphones can be considered creative improvements upon rotary and flip phones that required integration from all STEM fields with a foundation in scientific thinking. **Critical thinking** is applying intellectual standards to improve the quality of thinking (Paul, 1992; Elder & Paul, 2008). Coxon (2014) described critical thinking as a macroprocess necessary for the success of other processes. For example, to be successfully creative, one must be critical about which ideas to develop.

It is important to note that there is no more important aspect in student learning than the teacher (Chetty et al., 2011; Sanders & Horn, 1998). In discussions of 21st-century learning, technology is often seen as the heart of the conversation. It is ideal that students have access to and are taught to use relevant technology as tools to aid in practicing science; communicating, collaborating, and researching; and in the creation of products of their work to share. Ideally, quality science coursework for high-ability students integrates powerful curriculum in a Constructivist environment with the 4 Cs, modeling the work of professional scientists. However, there will be little benefit for students from either cutting-edge technology or well-constructed curriculum without the guidance of an excellent teacher. In the remainder of this chapter, curriculum and instructional methods based on this understanding are described, and programs, an example special school, and extracurricular programs fitting these dimensions are detailed. All are important, but excellent teachers are the key.

Instructional Methods

Based on this understanding of the ideal 21st-century science classroom, two of the strongest methods of secondary science instruction are project- and problem-based learning. Cotabish, Dailey, Coxon, Adams, and Miller (2014) provided concise descriptions of both project- and problem-based learning. Project-based learning can be described generally as any coursework that involves the creation of a product, whether it is a research poster or a digital story. Problem-based learning can be seen as a subset of project-based learning in which the final product is the result of students working toward solutions of a real-world problem. For teachers to appropriately carry out curriculum with these instructional methods, significant professional development should be undertaken. This is especially vital if teachers will be rewriting existing curriculum to be problem-based.

Project-Based Learning

Project-based learning in the secondary science classroom should be both based on real-world, current science and involve an audience for the product presentation. Both factors make the process more engaging for students. End products may be a scientific conference where papers or posters are presented to parents, younger students, or any other group that may have an interest. Student choice in what the final product will be is recommended both for motivation and, again, to parallel the work of 21st-century professional scientists. It is important from a practical standpoint that project-based learning incorporates metacognitive tools, such as graphic organizers for planning and deadlines. Students can also be involved in the development of rubrics and therefore be part of the assessment of their own outcomes. This may help with intrinsic motivation and much more closely simulates the critical nature by which scientists must observe their own work.

Problem-Based Learning

Problem-based learning is project-based learning in which students are given a real-world, ill-structured problem statement. Students work in collaborative teams and must determine what they know, what they need to know, and how they plan to find out the solution using resources like experts and processes such as the scientific method, critical thinking, and research. Students should often conduct experiments of their own design so that they are truly engaging in science as scientists. This is a big difference from other lab work, where students follow steps provided by the teacher. Studies involving problem-based learning units at the secondary level have demonstrated gains in student learning. VanTassel-Baska and Bass (1998) and VanTassel-Baska, Avery, Hughes, and Little (2000) both examined science units utilizing problem statements. Both found significant gains in student learning, particularly of the scientific process. Any science unit can be reorganized to be problem-based by creating a problem statement on the topic, providing students with some background knowledge, then providing sufficient time for them to conduct research, especially scientific experimentation, and to create final products demonstrating their learning. It is important for teachers to tailor the products and problems for student readiness, provide scaffolded learning, and provide structure such as small deadlines within larger projects. Several units exist, especially for the middle school level, including those created by the Center for Gifted Education and published by Kendall Hunt, such as *Acid, Acid Everywhere*, and *Animal Populations*. Such curriculum should be taught in appropriate contexts for the gifted.

> Any science unit can be reorganized to be problem-based by creating a problem statement on the topic, providing students with some background knowledge, then providing sufficient time for them to conduct research, especially scientific experimentation, and to create final products demonstrating their learning. It is important for teachers to tailor the products and problems for student readiness, provide scaffolded learning, and provide structure such as small deadlines within larger projects.

Programming Options

Advanced Coursework

Acceleration has a powerful effect on the academic achievement of high-ability learners (Colangelo, Assouline, & Gross, 2004; Rogers, 2007). As with mathematics, it is very important that students begin science study early and are allowed to progress at a pace suitable to their abilities. Novak's (2005) longitudinal studies over multiple decades clearly demonstrated that science instruction in elementary school results in better outcomes for high school seniors. Students that begin their science studies by second grade know more about science and hold fewer misconceptions as high school seniors than students who do not begin science education until middle school. Although nearly every U.S. elementary school claims to begin science education in early elementary school, it is my observation across dozens of districts in three states that science is rarely regularly and well taught by teachers with adequate science knowledge before middle school. Based on Novak's (2005) work, it seems likely that this lack of quality science instruction in elementary school has constrained the level of accomplishment that is generally attained by high school graduation and beyond. With that understanding, there are options for advanced science coursework available to many U.S. students.

Advanced Placement (AP) courses and the International Baccalaureate (IB) Program both offer advanced science options for secondary students, and their use with advanced students has been recently described by Hertberg-Davis and Callahan (2014). AP classes were developed as freshman college courses in the 1950s. AP science classes include physics, chemistry, biology, and environmental science. High school students take an end-of-course exam in which the scores are normed with college freshmen. AP courses are perhaps the most common form of advanced education offered in U.S. high schools: Almost two million high school students took an AP exam in 2011. The exams are moderately rigorous. Only 18% of those students earned a 3 or higher, the score most commonly accepted for credit at universities. The IB Program was developed in the 1980s to be a rigorous university preparation program (Hertberg-Davis & Callahan, 2014). IB Programs are offered at select schools and have expanded to include elementary and middle school programs. The high school program

> . . . lack of quality science instruction in elementary school has constrained the level of accomplishment that is generally attained by high school graduation and beyond.

leads to a special IB diploma. As of 2011, there were more than 2,200 IB high schools internationally.

The benefits of AP and IB for gifted and advanced learners are a subject of debate. Rogers' (2007) meta-analysis of 22 studies on AP and IB found only a small effect size ($ES = .29$). By comparison, 32 studies on grade skipping revealed a powerful effect size of 1.0. If AP courses were indeed accelerated by one year (typically, high school seniors taking college freshman-level courses), these effect sizes should be very similar. Although both students and teachers report being more satisfied with the courses and instruction in AP courses than with typical high school courses (Hertberg-Davis & Callahan, 2008; Hertberg-Davis, Callahan, & Kyburg, 2006), Scott, Tolson, and Lee (2010) found a significant yet meaningless effect size ($d = .08$) on student grades in the first semester at Texas A&M University for those who had taken an AP class in high school. AP, IB, and dual-enrollment are all associated with college success (Hertberg-Davis & Callahan, 2014), but correlation does not mean causation. It could be that such courses improve student success, or it could be that students who are already more likely to be successful (e.g., those of high ability, those from better high schools, those from high socio-economic status families) are more likely to participate in those programs in high school. Offering advanced students an accelerated IB pathway or AP classes earlier in high school may result in more meaningful effects on achievement. Where available, dual enrollment in strong, 4-year colleges and universities offers tremendous potential for advancing the academic achievement of high-ability students. Special schools offer the additional advantage of placing like-ability peers of the same age together.

Special Schools

Special schools have many advantages for the gifted: academically, socially, and emotionally (Coleman & Cross, 2005). Special schools that focus on STEM for bright students are rare, but ideal settings. When available at all, such schools are generally offered only at the high school level. They allow for high-ability students to pursue areas of strength and interest well before college, for which they traditionally must wait. If they were to become common, special STEM high schools may help to increase the number of college students pursuing STEM degrees.

A particularly apropos example is the Illinois Mathematics and Science Academy (IMSA; 2014), a public, competitive-entry residential school on the outskirts of Chicago. Originated by Nobel laureate Leon Lederman in 1986,

IMSA enrolls 650 students in grades 10–12. As recommended and described above, IMSA curriculum is project- and problem-based with extensive real-world connections and scheduled time for students to pursue their own academic interests. For example, Finley (2013) described how students developed a robotic arm during scheduled free time after learning that their principal had Lou Gehrig's Disease. Students participate in research with professional scientists and present at national conferences. The school boasts impressive outcomes: Nearly two-thirds of graduates go on to pursue STEM in college, compared with only 28% of college freshmen nationwide (Chen & Soldner, 2013; IMSA, 2014). Graduates have discovered new solar systems as well as helped to create companies such as PayPal and YouTube (Finley, 2013). IMSA demonstrates how the context of like-ability acceleration coupled with project- and problem-based learning in a Constructivist environment with excellent teachers can achieve maximized outcomes for high-ability secondary students in science. Extracurricular activities can enhance such ideal programs or be a partial substitute in their absence.

Extracurricular Programs

Science competitions, summer programs, and other similar opportunities can meet the academic and affective needs of high-ability secondary students when well selected. That is, they should involve excellent teachers focused on the students' areas of interest and strength. Ideally, extracurricular programs supplement excellent school programming, but too often such programs are all that are available to meet the needs of high-ability students when schools fail to offer accelerated and enriched science programming (Coxon, 2009). There are many benefits of extracurricular programs for high-ability students, including science talent development, opportunities for genuine mentorships, open-ended problems, higher order thinking, challenging tasks, no ceiling on excellence, and opportunities to work with ability peers (Omdal & Richards, 2008; Ozturk & Debelak, 2008).

Many universities and school districts offer science-focused afterschool and summer opportunities and a growing number of science competitions are available globally. Whenever possible, look for those aimed at gifted or high-ability students. For example, the Johns Hopkins Center for Talented Youth in Baltimore offers on-campus programs in a number of fields, including science as well as online science classes for students through grade 12 (see http://cty.jhu.edu/grade-by-grade/grades9-12/ for details). The Maryville Summer Science and Robotics Program in St. Louis offers around 80 classes for students through middle school in all areas of science, technology, engineering, art, and math (for details, visit https://www.maryville.edu/robot).

Competitions may increase creativity and motivation, improve self-concept, and help students to set higher goals (Omdal & Richards, 2008; Ozturk & Debelak, 2008). Many science-focused competitions are available to secondary students, including Science Olympiad, the Intel International Science and Engineering Fair, and Young Naturalist Awards. FIRST Robotics competitions engage students in grades 1–12 in all elements of STEM. In particular, the FIRST LEGO League served more than 400,000 students through middle school in 2014. The competition includes an annual real-world science problem related to what participants will engineer and program their robots to perform. For example, in a past competition, students learned about energy use, programmed their robot to place a LEGO solar panel on a LEGO house among other tasks, and could do a related science project, such as perform an energy audit on an actual building and research possibilities for improving energy use in that setting. Science Buddies maintains a list of science fairs and competitions at the following webpage: http://www.sciencebuddies.org/science-fair-projects/scifair.shtml.

Conclusion

It is important that more and better prepared high-ability students enter science fields. This will require that secondary science education be more engaging for students and authentic to professional science. Constructivist science classrooms that use project- and problem-based learning to teach science processes, content, and concepts may be best suited to accomplish this vital task when led by excellent teachers. Offering high-ability students advanced programming options, including earlier AP classes, special schools, and extracurricular programs, are all options in meeting the needs of gifted learners so that their talent development in science is maximized.

Key Recommendations

1. Focus resources on science, emphasizing attracting and retaining excellent teachers.
2. Offer high-ability students accelerated IB Programs or AP classes earlier in high school with options for dual enrollment in a university later in high school. Advocate for special STEM schools where they do not yet exist.
3. Provide significant professional development for teachers on project- and problem-based learning.
4. When implementing project- and problem-based learning, tailor the products and problems for student readiness; provide scaffolded learning; and provide structure, such as small deadlines within larger projects.
5. Seek out science-focused competitions and other extracurricular programs for high-ability students in your region.

Discussion Questions

1. Describe how your school science classrooms fit the chapter's description of Brooks and Brooks' (1993) 12-point framework of Constructivist classrooms. Identify and explain the importance of key areas for improvement.
2. If you are already implementing project- or problem-based learning in your classroom, describe your experiences and areas to focus on for improvement. If you are not already implementing either method, describe how you could begin using a specific unit of instruction.
3. What science-focused advanced coursework and extracurricular options are available to students in your area? What else could be done to aid the science talent development of students in your area? Prioritize the list and make a detailed recommendation to school policy makers.

References

Brooks, J. G., & Brooks, M. G. (1993). *The case for Constructivist classrooms.* Alexandria, VA: ASCD.

Chen, X., & Soldner, M. (2013). *STEM attrition: College students' paths into and out of STEM fields statistical analysis report.* Institute for Educational Statistics: National Center for Educational Statistics. Retrieved from http://nces.ed.gov/pubs2014/2014001rev.pdf

Chetty, R., Friedman, J. N., Hilger, N., Saez., E., Schanzenbach, D. W., & Yagan, D. (2011). How does your kindergarten classroom affect your earnings? Evidence from Project Star. *The Quarterly Journal of Economics, 126*(4), 1593–1660.

Colangelo, N., Assouline, S., & Gross, M. (Eds.). (2004). *A nation deceived: How schools hold back America's brightest students.* Iowa City: University of Iowa, The Connie Belin and Jacqueline N. Blank International Center for Gifted Education and Talent Development.

Coleman, L. J., & Cross, T. R. (2005). *Being gifted in school* (2nd ed.). Waco, TX: Prufrock Press.

Coxon, S. V. (2009). Challenging neglected spatially gifted students with FIRST LEGO League. In J. VanTassel-Baska (Ed.), *Addendum to leading change in gifted education.* Williamsburg, VA: Center for Gifted Education.

Coxon, S. V. (2014). Scientifically speaking: Nurturing student thinking isn't a frill: It's critical! *Teaching for High Potential, Summer,* 4.

Elder, L., & Paul, R. (2008). *Intellectual standards: The words that name them and the criteria that define them.* Berkeley, CA: The Foundation for Critical Thinking.

Finley, K. (2013, May 31). Hogwarts for hackers: Inside the science and tech school of tomorrow. *Wired.* Retrieved from http://www.wired.com/2013/05/hogwarts-for-hackers/

Hertberg-Davis, H., & Callahan, C. M. (2008). A narrow escape: Gifted students' perceptions of Advanced Placement and International Baccalaureate Programs. *Gifted Child Quarterly, 52,* 199–216.

Hertberg-Davis, H., & Callahan, C. M. (2014). Advanced Placement and International. Baccalaureate programs. In J. A. Plucker & C. M. Callahan (Eds.), *Critical issues and practices in gifted education* (pp. 47–64). Waco, TX: Prufrock Press.

Hertberg-Davis, H., Callahan, C. M., & Kyburg, R. M. (2006). *Advanced Placement and International Baccalaureate programs: A fit for gifted learners?* (RM06222). Storrs: University of Connecticut, The National Research Center on the Gifted and Talented.

Illinois Mathematics and Science Academy. (2014). *About IMSA.* Retrieved from https://www.imsa.edu/discover

Kim, K. H., & Coxon, S. V. (2013). The creativity crisis, possible causes, and what schools can do about it. In J. B. Jones. & L. J. Flint (Eds.), *The creative imperative: School librarians and teachers cultivating curiosity together* (pp. 53–70). Santa Barbara, CA: Libraries Unlimited.

Langdon, D, McKittrick, G., Beede, D., Khan, B., & Doms, M. (2011, July). *STEM: Good jobs now and for the future.* U.S. Department of Commerce Economics and Statistics Administration, Office of the Chief Economist, Issue Brief #03-11. Retrieved from http://www.esa.doc.gov/sites/default/files/reports/documents/stemfinalyjuly14_1.pdf

National Academy of Sciences. (2007). *Rising above the gathering storm.* Washington, D.C.: National Academy Press. Retrieved from http://www.nap.edu/catalog.php?record_id=11463

National Science Board. (2010). *Preparing the next generation of STEM innovators: Identifying and developing our nation's human capital.* Arlington, VA: National Science Foundation.

Novak, J. D. (2005). Results and implications of a 12-year longitudinal study of science concept learning. *Research in Science Education, 35*(1), 23–40.

Omdal, S. N., & Richards, M. R. E. (2008). Academic competitions. In Plucker, J. A. & Callahan, C. M. (Eds.), *Critcal issues and practices in gifted education,* (pp. 5–14). Waco, TX: Prufrock Press.

Ozturk, M. A., & Debelak, C. (2008). Affective benefits from academic competitions for middle school gifted students. *Gifted Child Today, 31*(2), 48–53.

Paul, R. (1992). *Critical thinking: What every person needs to survive in a rapidly changing world.* Berkeley, CA: The Foundation for Critical Thinking.

Rogers, K. B. (2007). Lessons learned about educating the gifted and talented: A synthesis of the research on educational practice. *Gifted Child Quarterly, 51*(4), 382–396.

Sanders, W. L., & Horn, S. P. (1998). Research findings from the Tennessee Value-Added Assessment System (TVAAS) Database: Implications for educational evaluation and research. *Journal of Personnel Evaluation in Education, 12*(3), 247–256.

Scott, T. P., Tolson, H., & Lee, Y-H. (2010). Assessment for Advanced Placement participation and university academic success in the first semester: Controlling for high school academic abilities. *Journal of College Admission, Summer, 208,* 27–30.

VanTassel-Baska, J. (1992). *Planning effective curriculum for gifted learners.* Denver, CO: Love Publishing.

VanTassel-Baska, J., Avery, L. D., Hughes, C. E., & Little, C. A. (2000). An evaluation of the implementation of curriculum innovation: The impact of William and Mary units on schools. *Journal for the Education of the Gifted, 23,* 244–272.

VanTassel-Baska, J., & Bass, G. (1998). A national study of science curriculum effectiveness with high ability learners. *Gifted Child Quarterly, 42*(4), 200–211.

CHAPTER 4

T is for Technology Education

Developing Technological Talent and Skill Through Curriculum and Practice

Angela M. Housand, Ph.D.,
& Brian C. Housand, Ph.D.

Introduction

For a century, many in the United States took for granted that most great inventions would be homegrown—such as electric power, the telephone, the automobile, and the airplane—and would be commercialized here as well. But we are less certain today who will create the next generation of innovations, or even what they will be. We know that we need a more secure Internet, more-efficient transportation, new cures for disease, and clean, affordable, and reliable sources of energy. But who will dream them up, who will get the jobs they create, and who will profit from them? (National Research Council [NRC], 2007, p. 40)

Technology, the "T" in STEM is the means by which a more secure Internet, more efficient transportation, new cures for disease, and new energy sources are achieved—but what is technology? What is its relationship to the other STEM areas, and how is technology education addressed in K–12 settings?

 DOI: 10.4324/9781003238218-5

The National Research Council (2007) characterized technology as the entire system of people and organizations, knowledge, processes, and devices that go into using and creating technological artifacts. Technology is "the innovation, change or modification of the natural environment in order to satisfy perceived human wants and needs" (International Technology Educators Association [ITEA], 2007, p. 242). In other words, technology exists to modify the world to meet human needs. It is difficult, however, to discuss technology in STEM without also understanding its relationship to science, engineering, and mathematics, as technology is unique from the other STEM fields, but would not exist without them. The International Technology Educators Association (2007) clarified the relationships:

> ▷ Science is the foundation of technology that seeks to understand the world as it currently exists, whereas technology seeks to define the world through invention, innovation, and design. Technology addresses what could be rather than what is.
> ▷ Mathematics, the systematic treatment of relationships and magnitudes, provides the language for technology, science, and engineering. Technology and mathematics share a reciprocal relationship as advances in either area inform and advance the other.
> ▷ Engineering, maybe most closely related to technology, is the development of practical application for mathematics and science knowledge to benefit humankind. Like technology, engineering has aspects of problem solving and innovation using science and mathematics.

It might be said that science and mathematics provide the foundational knowledge and skills necessary for advancement in the fields of technology and engineering, and technology serves to illustrate the outcomes of engineering design.

Integration

Students who study technology learn about a world created by engineers and innovators. Fields such as energy production, communication, manufacturing, chemical engineering, computer sciences, transportation, and medicine all use technologies created as a result of engineering processes. Because technology is so integrated into all of these fields, those who teach technology tend to focus on concepts and principles rather than specific details (ITEA, 2007).

This concepts-and-principles view of instruction is not new to gifted education. For decades, curriculum experts have been espousing an approach to curriculum that focuses on broad concepts and principles, as well as illustrating the importance of curriculum that is integrative and focusing on connections between domains of knowledge and across disciplines (Kaplan, 1986; Renzulli, Leppien, & Hays, 2000; Renzulli & Reis, 1997; Tomlinson et al., 2009; VanTassel-Baska, 2011). An integrative curriculum focusing on connections between disciplines allows students to find different points of entry, connect their interests to the academic content, and combine information from a variety of fields to develop novel and insightful solutions to problems they encounter within sufficiently advanced learning opportunities. These experts also seem to agree that curriculum is relevant for students when it connects with their lives, seems useful in contexts beyond the classroom, allows for meaningful collaboration, is sensitive to global concerns, and is authentic, focuses on real problems and processes, uses the conventions of the discipline, and guides by habits of mind (Eccles & Wigfield, 1995; Kaplan, 1986; Renzulli, Leppien, & Hays, 2000; Renzulli & Reis, 1997; Tomlinson et al., 2009; VanTassel-Baska, 2011). Successful STEM education seems to be on the right track, as it reflects similar curriculum priorities that have been championed by the field of gifted education.

Take, for example, extracurricular learning opportunities in a STEM-focused charter school. Sahin, Ayar, and Adiguzel (2014) tracked 146 students in grades 4–12 participating in a variety of self-selected afterschool programs. Out of the 146 students, 17 participated in the robotics afterschool program and 33 did science-related programs (both of these presented integrated learning opportunities), with another 15 focusing on mathematics. STEM-related afterschool program activities increased students' interest in participating and pursuing STEM. In addition to increased interest in STEM, students considered the activities appealing because they were comprehensive (e.g., science, math, and engineering required to design robotics technology) and they were fully engaged in the activities, reporting feeling more productive, successful, and happy. Through the integration of multiple domains, students found relevance, importance, and enjoyed the process of learning, so implementing a technology and engineering integrated curriculum should be easy, right?

Challenges to Integration

We do face challenges in our efforts to implement curriculum that is integrative, develops technology talent, and increases technology literacy. One challenge is that historically core curriculum areas have been mathematics, English language arts, science, and social studies, yet the spectrum of domains has been eroded in recent years with math and reading replacing time spent on science, social studies, and the arts (McMurrer, 2008). Ideally, the other areas would be subsumed by reading and mathematics, but it is more likely that more time is being given to drill and practice in these domains to ensure that students are all performing at grade level or better. This then leaves little room or opportunity in the instructional day for focusing on the core elements of engineering and technological advancement such as design, development and production, and testing the limits of a product (ITEA, 2007).

Another challenge is the current focus in classrooms on test performance over innovation. We cannot deny the reality that we are living in an educational climate that is governed by accountability and is guided by standards. Although these are in service to improving education, we cannot forgo innovation and enjoyment in learning for the sole purpose of improving test scores. Otherwise, outside of classrooms, technology will continue to advance at an unprecedented rate and the United States will lose its entrepreneurial edge.

The International Society for Technology in Education (ISTE) has been very progressive in how to approach technology use in schools. ISTE developed standards that address the need for educational accountability, while honoring the ever-changing technological landscape. The ISTE Standards (2007) incorporated many of the principles and best practices of gifted education with technology. Standards have been developed for students, teachers, administrators, coaches, and computer science educators outlining the essential skills necessary to demonstrate technology literacy—an important goal put forth by both the ITEA (2007) and the NRC (2007, 2009, 2012). Rather than being subject specific, the ISTE Standards for students offer a cross-curricular view of the necessary skills for students to be successful and competitive in a globalized and technologically advanced world. The ISTE Standards for students are divided into six separate strands that in many ways echo the exact principles that many gifted programs are based on:

> ▷ Creativity and Innovation;
> ▷ Communication and Collaboration;
> ▷ Research and Information Fluency;

> ▷ Critical Thinking, Problem Solving, and Decision Making;
> ▷ Digital Citizenship; and
> ▷ Technology Operations and Concept.

It is important here that clarification be made about a few things: First, technology education and educational technology (i.e., instructional technology) are often used interchangeably; however, they are not the same thing. Educational technologies are technology tools like computers, audiovisual equipment, graphing calculators, and software that are used to support instruction, deliver curriculum, assess learning, or otherwise enhance teaching and learning environments (ITEA, 2007), whereas technology education is the study of technology. Second, the ISTE Standards (2007) are not technology education specific, but rather address the need for technology and learning to fuse seamlessly so that basic technological skill and literacy is fundamental to all learning processes. Rather than addressing technology as something separate, the ISTE Standards for students provide the guidance that enables students to build a strong foundation for more advanced exploration of technology skills, innovation in technology fields, and ensures the integration of technology across domains of knowledge and skill areas.

> . . . in the same way that the ISTE Standards have similarities to gifted pedagogy (e.g. creativity, critical thinking, problem solving) but are not themselves pedagogy, STEM approaches may have similarities to gifted education, but STEM education will not, does not, and should not replace gifted education.

Finally, in the same way that the ISTE Standards have similarities to gifted pedagogy (e.g. creativity, critical thinking, problem solving) but are not themselves pedagogy, STEM approaches may have similarities to gifted education, but STEM education will not, does not, and should not replace gifted education. In our efforts to develop talent, we must remember that talented students provide the promise of a better future but that an individual's wish for his or her own trajectory should take precedence. How, then, do we make sure that students fully understand what education in technology and STEM have to offer, and how do we ensure that we create a stream of STEM talent?

Creating a STEM Talent Stream

There have been numerous calls for advancing the STEM agenda by developing the talents of our best and brightest (e.g., NRC, 2007; National Science

Foundation [NSF], 2010; Gallagher, 2013). The report by the National Science Foundation (2010), *Preparing the Next Generation of STEM Innovators*, focused specifically on identifying and developing talent in STEM. The report recommended providing opportunities for excellence, casting a wide net to identify talent, and fostering environments that nurture and celebrate excellence and innovation. One way to achieve all three of these goals exists in the current structure of our educational setting, but we have to think differently about how we approach instruction, and we have to start exposing students to technology and engineering concepts at the beginning of their educational careers.

Recall that math and reading have replaced time spent on science (McMurrer, 2008); however, it is not those who have high reading and math scores that do well in STEM, but rather those that have high-spatial ability and math attainment. Wai, Lubinski, and Benbow (2009), using data from the Project TALENT longitudinal study, found that the likelihood of earning a STEM degree is related to spatial ability and that the likelihood of earning an advanced degree in STEM increases as spatial ability increases. Unfortunately, 70% of the students talented in spatial ability (i.e., the top 1%) are not in the top 1% of tests of verbal and math ability (Wai, Lubinski, & Benbow, 2009). Coupled with the fact that the representation of economically disadvantaged, English language learners, and historically underprivileged minorities within the top levels of achievement on standardized tests is proportionally smaller than other groups (Plucker, Burroughs, & Song, 2010), it is likely that many of the talent search programs miss potential talent in the STEM fields because they restrict their identification practices to mathematical and verbal ability. Moreover, spatial ability is seldom developed in K–12 classrooms because it is rarely included as part of the benchmark assessments that tend to drive instructional practice in the current educational zeitgeist, yet spatial ability reliably predicts participation in STEM fields in adulthood (Shea, Lubinski, & Benbow, 2001; Webb, Lubinski, & Benbow, 2007). Therefore, we must value, nurture, develop, and identify spatial abilities in American classrooms to ensure a talent stream for technological innovation and the advancement of society.

> The [National Science Foundation] report recommended providing opportunities for excellence, casting a wide net to identify talent, and fostering environments that nurture and celebrate excellence and innovation. One way to achieve all three of these goals exists in the current structure of our educational setting, but we have to think differently about how we approach instruction, and we have to start exposing students to technology and engineering concepts at the beginning of their educational careers (NSF, 2010).

Developing Spatial Ability

It has been suggested that robotics may be an effective way to develop spatial talent in classroom settings (e.g., Coxon, 2009; Rockland et al., 2010; Verner, 2004), and research demonstrates that using robotics can increase spatial ability (Coxon, 2012; Verner, 2004). For example, Verner (2004) studied the effects of programming robot movements on spatial ability. Seventh graders (n = 61) were provided a 12-hour course on the spatial motion of robots. Precourse and postcourse assessments were used. The results showed significant improvements in the spatial skills of perception and visualization. Verner repeated the study multiple times with both middle and high school students. The high school students were given 22 hours of course instruction and showed significant improvements in the spatial skills of rotation and visualization.

Similarly, in a study using a simulation of the FIRST (For Inspiration and Recognition of Science and Technology) LEGO League competition, Coxon (2012) demonstrated meaningful gains in spatial ability. Using a randomized intervention design, 75 gifted students aged 9–14 were divided into two groups. After only 20 hours of participation in the FIRST LEGO League competition, males who participated in the intervention showed significant gains in spatial ability.

As Coxon (2012) and Verner (2004) have demonstrated, the competitions offered by FIRST can help to meaningfully engage students in challenges that incorporate technology while developing visual spatial ability. Begun in 1998, FIRST has been sponsoring an annual challenge to students of all ages known as the FIRST LEGO League (FLL). Teams consist of up to 10 students ages 9–16 and one adult coach. More than just a robotics competition, the FLL is comprised of three distinct but interconnected components. First, the robot game challenges teams to program an autonomous robot using LEGO MINDSTORMS technology to score points on a themed playing field. Second, the project asks teams to develop a solution to a problem that has been identified. Each fall, a new problem is introduced for the year. Past problems have included such topics as Nature's Fury, Senior Solutions, and Nano Quest. The 2014 problem had teams of students investigating what education may look like in the future. Although the first two components of the FLL focus on what teams will "do," the FLL Core Values focus on "how" the students will do it.

The FLL Core Values emphasize the importance of teamwork and the idea that individuals seldom solve problems as effectively as a team. This type of learning is precisely the type of thing that is missing in many high-stakes testing environments that our gifted students have grown accustomed. Additionally, the FLL Core Values point out that coaches and mentors do not have all of the answers, but instead we are all learning together. Although the FLL is in essence a competition, they choose to identify themselves as a Coopertition® (part cooperation, part competition), celebrating the idea that what students discover is more important than what they win.

Regardless of the grade level, FIRST has an event for students. Beginning with grades K–3, there is the Junior FIRST LEGO League, which centers on capturing the curiosity of young children and directing it toward discovery of science and technology concepts. In addition, there is the FIRST LEGO League for students in grades 4–8. FIRST also offers challenges for middle and high school students. For students in grades 7–12, there is the FIRST Tech Challenge. This differs from the previous two events in that it uses a sports model to have teams construct robots to compete in an alliance against other teams. Finally, the FIRST Robotics Competition for students in grades 9–12 challenges teams of 25 students to fundraise, design a team brand, and design and program robots using strict rules, limited resources, and time constraints. Additional information can be found at http://www.usfirst.org/.

Noticeably, the FIRST LEGO League has similarities to Renzulli's (1976) Enrichment Triad Model, in that the initial stages of the process begin with exposure experiences similar to those that might be included in a Type I. The second stage of FIRST LEGO League focuses on developing the knowledge base and skills to move to higher levels of investigation (not unlike a Type II), and once students are fully invested they are given the opportunity to engage in all aspects of professional technological challenges to innovate solutions to an authentic problem. This type of deep exploration of a topic is not unlike a Type III. Although the similarities are of note, these STEM educational opportunities are not the Enrichment Triad Model.

The Enrichment Triad Model

In 1976, Renzulli introduced the Enrichment Triad Model, a seminal work in the field of gifted education. The model seeks to encourage creative productivity through exposure, skill development, advanced exploration, application, and productivity for individuals or small groups. This Enrichment Triad Model goes well beyond the bounds of the FLL in that a primary goal of the

Enrichment Triad Model is for students to explore topics in their area of interest via three components known as Type I, Type II, and Type III.

The Type I component often serves as the initiation for Type II process skill development and deeper explorations of the Type III. In Type I experiences, students are exposed to a variety of topics, persons, places, events, and professional activities. Type I exposure experiences can be achieved in a variety of ways and may serve as a means to increase interest in STEM fields. For school settings, Renzulli and Reis (1997) suggested guest speakers, minicourses, demonstrations, performances, videos, or print media—just to name a few.

Type II experiences are enrichment opportunities that cannot be planned in advance because they are designed to be responsive to students' interests (potentially identified by participation in a Type I enrichment experience). Type II experiences provide instruction in the skills necessary for exploration of advanced content. Type II experiences might be considered the "how to" experience as students employ creative thinking and problems-solving skills to address authentic problems, utilize a wide variety of materials and resources, and communicate their learning via written, oral, or visual means.

Type III enrichment is an opportunity for students to examine a self-selected topic of interest in a way that requires the commitment of time, deep review and understanding of advanced content, the development of authentic products, the use of advanced skills like those used by professionals in students' domain of interest, as well as the skills of innovation, self-regulation, resource management, decision making, reflection, and evaluation.

The Enrichment Triad Model is ideally suited for use in the classroom to recognize students interested in technology and engineering. The model provides the framework to support students in the development of technology skills specifically and STEM skills more generally. Further, Type I, Type II, and Type III experiences require students to innovate, develop, produce, and assess by exposing students to a variety of topics, areas of interest, and fields of study. These experiences provide opportunities for the explicit instruction of "how-to" skills that will enable students to apply advanced content, and they allow students self-select (individually or in small groups) and deeply engage in authentic problem solving by assuming the role of first-hand inquirer. So, how do we leverage the Enrichment Triad Model to help students develop technology skills in classroom settings?

Type I: Exploratory Activities

With the advent of the Internet age, Renzulli's Type I experiences have been transformed. Once depending on physical access to speakers and a variety of tangible resources, Type I experiences required significant time to gain access to the people and resources that resulted in meaningful exposure experiences. Now, quality online resources and content are only a Google search away. Technology-related exposure experiences can be as simple as a TED Talk, TED ED mini course, videos via YouTube, a Skype guest speaker, a demonstration of cutting-edge technologies, a virtual field trip to a local university's high-tech laboratory, online courses through outlets like Coursera or iTunes U, or a mini lesson designed to simulate the use and development of technology.

One of the challenges of ubiquitous access to the Internet and all that it has to offer is that students are often lost when it comes to locating quality resources. According to a 2012 Pew Research Internet survey of more than 2,000 middle and high school teachers who participated in the National Writing Project or teach Advanced Placement courses, 83% responded that today's digital technologies discourage students from using a wide range of sources when conducting research (Purcell, et al., 2012). Therefore, instead of turning students to the virtual wilderness of Google—the digital equivalent of dropping students off at the Library of Congress—educators must begin to curate quality content sources for students.

Renzulli Learning, a learning analytics program that assesses students' interests, learning styles, and expression styles to guide the exploration of numerous online resources, is an example of automated curation. Renzulli Learning is a powerful tool that enables students to explore areas of interest. An illustration of the usefulness of resource curation comes from a study (Housand, 2008) reviewing the use and effects of Renzulli Learning. Over a series of observations, one student was first seen to be enthralled with a virtual roller coaster activity where students use different track templates to design a roller coaster. The site challenges students to make the fastest coaster with the greatest g-force. Students are then able to virtually ride their creations. The observed student became fascinated with the site and then wanted to share the site with the entire class. During the next observation, the same student found an activity, Tinker Ball, from the Lemelson Center for the Study of Invention and Innovation (http://inventionatplay.org/playhouse_tinker.html). The object of the activity is to create a path using various gears, tubes, boards, and other contraptions that would allow a ball to be released from a platform and land in a cup below. The student was observed working for more than 45 minutes devising different arrangements and increasingly complex designs to accomplish this task. Within this same study, another student was observed over multiple sessions researching everything available regarding rockets and rocket launches. Both these examples illustrate the exposure experience and how these Type I experiences can move students into deeper learning on topics in their area of interest. More specifically, these examples provide evidence of how these opportunities can be used by teachers to leverage students' intense enthusiasm and obvious interest in the technology of engineering or rockets to encourage a deeper exploration by creating an authentic product in their area of interest.

The Internet increases accessibility on a global scale and it is a fact of life that is not changing anytime soon. No longer is technology merely a tool, it is the medium for attaining knowledge, collaborating with peers, exchanging ideas, creating products, and sharing knowledge and insights. This kind of access also provides infinite opportunities for exposure to STEM fields, technology MOOCs, simulation activities, and opportunities to access communities and mentors that can provide access to not only the knowledge base for STEM learning, but also the social interactions that can lead to increasingly complex and meaningful STEM learning opportunities. Technology tools and access to the Internet are fundamental to the learning process, particularly when the very goal of instruction is to develop technology skills, but it is important to be aware that students today are adept at socializing via the Internet, but may lack the skills of critical consumerism and digital citizenship.

Type II: "How To" Skill Training

In Renzulli's Three Ring Conception of Giftedness (1978), he emphasized the important distinction between consumers of information and producers of new knowledge. The authors of this chapter would also like to emphasize the importance of not only having students consume or use technology, but the possibilities of having talented students learn how to produce their own programs and design their own hardware, as well. It is in these endeavors that technology talent can be developed and fully realized. However, before students can begin this transformation, they must be provided with some instruction on how to accomplish this. An important first step in this journey from consumer to producer may come with an introduction to computer programming or coding.

Computer science and programming. Based on the 2012 report from the Bureau of Labor Statistics, Code.org (2015; http://code.org) estimated that by the year 2020, there will be one million more jobs in the field of computer sci-

[We would] like to emphasize the importance of not only having students consume or use technology, but the possibilities of having talented students learn how to produce their own programs and design their own hardware, as well. It is in these endeavors that technology talent can be developed and fully realized. However, before students can begin this transformation, they must be provided with some instruction on how to accomplish this. An important first step in this journey from consumer to producer may come with an introduction to computer programming or coding.

ence than there are students in computer science programs. Launched in 2013, Code.org seeks to expand interest and participation in the burgeoning field of computer science by making introductory courses available to students and teachers for free.

Code.org offers a wide range of tutorials and learning modules for students of all ages. The Hour of Code was designed as a tutorial for beginners to learn some of the basic concepts of Computer Science via a "drag and drop" programming tool. In the introductory course, students view video mini lectures from Bill Gates, Mark Zuckerberg, and many other well-known computer scientists and then complete a series of game-based levels while learning about repeat loops, conditional statements, and other basic algorithms. Beyond the Hour of Code, Code.org offers tutorials to teach programming languages such as Java Script and Python, resources for making your own apps, and tutorials on creating your own games. Additional learning modules and courses continue to be developed and added to the Code Studio (http://studio.code.org/). The site currently features three 20-hour courses designed for elementary students.

Although Code.org tends to focus on introducing students to computer science, Code Academy (http://www.codecademy.com/) offers a logical next step for students who are ready to learn core programming skills for popular and widely used languages, including HTML, CSS, PHP, Python, and Ruby. Set up as a simulation, users work their way through a set of increasingly difficult challenges to acquire the knowledge and skillset of each language. Code Academy also offers a collection of mini projects for students to put their newly acquired skills to work in a real-world type setting.

Research supports this type of skill development, using coding to develop a web-based game. Reynolds and Caperton (2011), in an educational pilot study, engaged 199 students from middle school to community college in a learning opportunity that required each student to actively develop an online game. As part of the learning experience, students were enrolled in a game design class, had access to a wiki-based e-learning environment, had access to game design professionals via Skype, and shared their learning in a community space for the course. As part of the course, students had to learn programming languages, use computational design tools, and create complex representational digital arti-facts. In other words, they had to learn the skills that would enable them to suc-cessfully complete the process of designing a web-based game. Once again, this example also illustrates how students utilized transferable, authentic skills by engaging in game design in technological formats. Some of the "authentic and transferable" skills identified by students included learning teamwork, people skills, or social skills; proposing a project and following it through to comple-tion; as well as perseverance and patience. Their perception of the learning that was occurring differed dramatically from that which they experienced in their more traditional educational settings. Students suggested that their experience in designing games made learning fun and not boring; it provided opportuni-ties to engage in self-directed learning, teamwork, and cooperation; the learn-ing environment was more relaxed and less pressured; they got to experience hands-on learning; they learned "new" things; and the work itself was hard and challenging, but also interesting.

Hardware. Although learning to code a computer and create your own content and apps may be the first step toward creative productive giftedness in the technological domain, there exists something beyond the user interface that also is compellingly interesting to technologically talented students. For these students, the machine itself motivates them. Just as a growing number of coding resources for students have emerged, a number of hardware devices have been developed as well. Initiated by two students at the MIT Media Lab, the Makey Makey Project (http://makeymakey.com/) is described as "an invention kit for everyone." The kit costs $49.99 and includes a computer circuit board with a directional keypad and assorted buttons, as well as connector wires and alligator clips. The kit allows the user to connect the circuit board to almost any physical object to control the associated keys on a connected computer. Project ideas include creating an electric piano made from bananas and a video game control-ler made from Play-Doh. This type of device can serve to spark the creativity and imagination of talented students to further explore computer hardware devices.

Another option for introducing students to computer hardware is littleBits (http://littlebits.cc). More than 60 interchangeable electronic modules are available for purchase in a variety of different kits and collections. The modules are designed to connect together to allow students to devise their own elec-tronic circuits by clicking individual pieces together. Once assembled, the new device is capable of gathering input from a variety of environmental data and provides a wide range of output options. littleBits also offers a wealth of educa-tor resources and lesson ideas for teaching students how to imagine and create new computer devices (http://littlebits.cc/education).

A next logical step in this development is the Raspberry Pi (http://www. raspberrypi.org/). Available for $35, the Raspberry Pi is a credit card-sized mini computer that can plug into a TV and USB keyboard. The device is capable of being programmed to do almost anything that a typical PC might be able to do. However, the relatively low cost and open architecture encourages students to create new types of devices and uses for this computer. As an added bonus, teaching resources and instructional units have already been developed for using the Raspberry Pi to teach students how to set up their own hardware.

> As students learn the "how to" skills of computer programming and begin exploring the possibilities of using a number of introductory hardware devices like the ones listed, natural curiosity leads them from this Type II skill development to the world of Renzulli's Type III.

Finally, Arduino (http://www.arduino.cc/) is an open-source computing platform that can be used to develop interactive objects that accepts input from a variety of different sensors and switches and is capable of producing a variety of outputs, including control of lights, motors, and other devices. A variety of different Arduino kits are available for purchase.

As students learn the "how to" skills of computer programming and begin exploring the possibilities of using a number of introductory hardware devices like the ones listed above, natural curiosity leads them from this Type II skill development to the world of Renzulli's Type III.

Type III: Investigations of Real Problems

In Renzulli's Enrichment Triad, Type III involves students pursuing self-selected topics of interest and developing authentic products that require task commitment, application of content knowledge, an understanding of methodology, and creativity. Each of the resources related to computer programming and hardware development mentioned previously has the potential to evolve into a Type III project. Providing the opportunity, resources, and encouragement allows students to find advanced problems and develop their own products in independent or small-group investigations.

In a sense, the very crux of technological advancement is creativity and innovation, but these have little benefit if not combined with a productivity component. Type III require students to develop authentic products. If their interests rest in STEM, they are well on their way, but what are some things that might be useful in a classroom setting?

The Maker Movement is an emerging and increasingly popular phenomenon that is finding its way from the world of hobbyists and amateur inventors into educational environments. In 2005, Dale Dougherty launched *Make* magazine (http://makezine.com/) as a bimonthly publication focusing on "do it yourself" (DIY) projects involving technology, electronics, robotics, and various construction-related projects. As an increasing number of individuals and small groups developed new products and creations, they began seeking out opportunities to learn from and share with one another. In 2006, this need gave rise to Maker Faires, which have become regular events held around the world. The 2012 Maker Faire in San Mateo, CA, was visited by more than 120,000 people interested in learning firsthand more about robots, 3D printing, and other DIY creations.

As the Maker Movement gains popularity, community centers and many public libraries have begun establishing Makerspaces for inventors of all ages to come together to imagine, build, and create. According to the Makerspace Playbook (http://spaces.makerspace.com/), "Makerspaces represent the democratization of design, engineering, fabrication, and education" (para. 2). An increasing number of schools are exploring ways they can incorporate Makerspaces into the learning environment.

Another technology on the horizon that could serve to support student creativity, design, and innovation is additive manufacturing (AM), also known as 3D printing. 3D printing is process whereby three-dimensional solids are made from a digital file. Objects are created as successive layers of material are built up in a repeated fashion. AM technologies are notably suited for product development, data visualization, and rapid prototyping. In recent decades, strides have been made to expand the use to mass production and distributed manufacturing. The use of 3D printing continues to evolve. For example, exciting developments in the field include 3D bioprinting, which uses cells and encapsulation material to print several kinds of tissue structures such as skin, bone, cartilage, trachea, and heart (Murphy & Atala, 2014). The uses of 3D printing are numerous: bioprinting, nanoscale printing, apparel production, automobile prototyping, construction, motors and generators, firearms, and art and communication. 3D printing itself is an area for Type I and II exploration, but the use of a 3D printer in classroom settings holds significant promise for encouraging innovation, supporting design processes, developing and producing finished products, and even testing products to determine their success or failure. In other words, 3D printing is a meaningful tool to enable students to engage in the basic elements of technology and engineering.

Although 3D printers are currently cost prohibitive, like any rapidly advancing technology, the price is already dropping to within reasonable ranges. A simple Google search leads to examples of personal 3D printers for a mere $999. So, as the price drops, the future of classroom printers must necessarily change, because the products we ask students to create are not just about today, but also about tomorrow. We are preparing technologically talented young thinkers to develop as yet unimagined FUTURE products.

Conclusion

By far, one of the most important aspects of our educational system is the teaching staff. Teachers have tremendous power to influence students in their classrooms, who will go on to be future workers. Teachers can inspire students to pursue STEM careers and they can provide appropriate curricular options for students to develop their potential in the design, development, and production of technologies that will solve some of society's most pressing problems. Talking about developing talent in technology without addressing the need for high-quality teachers would be an oversight.

The importance of teachers is clearly recognized by national experts. Of the 20 action steps recommended by the NRC in 2007, the first two specifically and directly addressed teachers. Take, for example, the action recommendation, a quarter of a million teachers inspiring young minds every day. This recommendation had four parts, the first three of which could be summarized as strengthening the skills of 250,000 teachers through training and education programs at summer institutes, in master's programs, and in Advanced Placement (AP) and International Baccalaureate (IB) training programs.

What the report seemed to miss was the need for preparing teachers to develop talent, as talent development plays an important role in developing technology talent. Researchers in gifted education have highlighted the need for preparing teachers to serve students with gifts and talents, and several suggest that all educators should receive instruction, training, or professional development about the characteristics of gifted students and meeting the needs of this unique population (Cramer, 1991; Feldhusen, 1997; Gallagher, 2000). Researchers have also shown that teachers who have training in gifted education make curricular modifications for gifted students, demonstrate greater teaching skills, and develop more positive class climates (Archambault et al., 1993; Hansen & Feldhusen, 1994; Westberg & Daoust, 2003; Whitton, 1993).

For example, Westberg and Daoust (2003), in a replication study of Archambault et al. (1993), found significant differences in curricular modifications for gifted students between teachers who had taken coursework in gifted education and those who did not. Further, those teachers who had earned formal degrees in gifted education provided challenges and choices to all of their students and curriculum modifications for high-ability students. Of note is the fact that having earned a formal university degree reinforces the idea that all students need to be challenged and provided the opportunity for choice. The formal degree also seems to reinforce the fact that gifted students need something *different* and more advanced than their grade-level peers to ensure sufficiently challenging curricular options that support continuous growth for even the most able students.

Developing the talents of all students means valuing a variety of perspectives and abilities and guiding them to a mutually beneficial goal. This is illustrated by Project Hieroglyph, a collaborative project produced by Arizona State University's Center for Science and the Imagination. The project was designed to "rekindle grand technological ambition" (Finn & Cramer, 2014, p. 265) and highlight the role of science fiction and science fiction writers in conjuring up ambitious futures that inspire engineers to make the imagined real through innovation and technological advances. Those who promote talent development in STEM must seek to go beyond textbook-style instruction, consider the many and varied ways that talent manifests, and be willing to take risks in their instructional settings to set a new course for students' futures and for the futures of every individual who might benefit from technological innovation—which is everyone.

We are preparing students for jobs that do not yet exist using technologies that have yet to be imagined. We have to move beyond a 21st-century skills mindset and seek a "change and innovation" mindset, and we must embrace the notion that technology is evolving at an alarmingly fast pace and the projection of that pace will inevitably continue to increase. To be competitive now and in the future, individuals will have to take more initiative, be more responsible, and produce more than ever before. They will have to be flexible, comfortable with ambiguity, and continually create and recreate to stay viable in a world that will, evermore, be in a state of flux. They will have to go where no one has gone before . . .

Discussion Questions

1. Why is technological advancement important for the advancement of society?
2. Why do those who teach technology tend to focus on concepts and principles? What steps might you take to focus instruction in the classroom on concepts and principles?
3. How might you increase students' interest for participating in and pursuing STEM-related careers?
4. How might you develop spatial ability in classroom settings? What kinds of curriculum could help support the development of spatial ability?
5. How might you use the Enrichment Triad Model to help students develop technology skills in classroom settings?

References

Archambault, F. X., Jr., Westberg, K. L., Brown, S. W., Hallmark, B. W., Emmons, C. L., & Zhang, W. (1993). *Regular classroom practices with gifted students: Results of a national survey of classroom teachers* (Research Monograph 93102). Storrs: University of Connecticut, The National Research Center on the Gifted and Talented.

Code.org. (2015). Promote computer science. Retrieved from http://code.org/promote

Coxon, S. V. (2009). *Challenging neglected spatially gifted students with FIRST LEGO League. Addendum to leading change in gifted education.* Williamsburg, VA: Center for Gifted Education.

Coxon, S. V. (2012). The malleability of spatial ability under treatment of a FIRST LEGO League-based robotics simulation. *Journal for the Education of the Gifted, 35,* 291–316.

Cramer, R. H. (1991). The education of gifted children in the United States: A Delphi study. *Gifted Child Quarterly, 35,* 84–91.

Eccles, J., & Wigfield, A. (1995). In the mind of the actor: The structure of adolescents' achievement task values and expectancy-related beliefs. *Personality and Social Psychology Bulletin, 21*(3), 215–225.

Feldhusen, J. (1997). Educating teachers for work with talented youth. In N. Colangelo & G. Davis (Eds.), *Handbook of gifted education* (pp. 547–552). Boston: Allyn & Bacon.

Finn, E., & Cramer, K. (Eds.) (2014). *Hieroglyph: Stories and visions for a better future.* New York, NY: HarperCollins.

Gallagher, J. J. (2000). Unthinkable thoughts: Education of gifted students. *Gifted Child Quarterly, 44,* 5–12.

Gallagher, J. J. (2013). Educational disarmament, and how to stop it. *Roeper Review, 35,* 197–204.

Hansen, J. B., & Feldhusen, J. F. (1994). Comparison of trained and untrained teachers of gifted students. *Gifted Child Quarterly, 38,* 115–121.

Housand, B. C. (2008). *The effects of using Renzulli Learning on student achievement and motivation.* Unpublished doctoral dissertation, University of Connecticut, Storrs.

International Society for Technology in Education. (2007). *ISTE standards.* Washington, DC: Author. Retrieved from http://www.iste.org/standards.

International Technology Educators Association. (2007). *Standards for technological literacy: Content for the study of technology* (3rd ed.). Reston, VA: Author.

Kaplan, S. N. (1986). The grid: A model to construct differentiated curriculum for the gifted. In J. S. Renzulli (Ed.), *Systems and models for developing programs for the gifted and talented* (pp. 180–193). Waco, TX: Prufrock Press.

McMurrer, J. (2008). *Instructional time in elementary schools: A closer look at changes for specific subjects.* Washington, DC: Center for Education Policy.

Murphy, S., & Atala, A. (2014). 3D bioprinting of tissues and organs. *Nature Biotechnology, 32*(8), 773–785.

National Research Council. (2007). *Rising above the gathering storm: Energizing and employing America for a brighter economic future.* Washington, DC: The National Academies Press. Retrieved from http://www.nap.edu/catalog.php?record_id=11463

National Research Council. (2009). *Engineering in K–12 education: Understanding the status and improving the prospects.* Washington, DC: The National Academies Press.

National Research Council. (2011). *A framework for K–12 science education: Practices, crosscutting concepts, and core ideas.* Washington, DC: The National Academies Press.

National Science Foundation. (2010, May). *Preparing the next generation of STEM innovators: Identifying and developing our nation's human capital*

(NSB-10-33). Arlington, VA: National Science Board. Retrieved from http://www.nsf.gov/nsb/publications/2010/nsb1033.pdf

Plucker, J. A., Burroughs, N., Song, R. (2010). *Mind the (other) gap! The growing excellence gap in K–12 education.* Bloomington, IN: Center for Evaluation & Education Policy.

Purcell, K., Rainie, L., Heaps, A., Buchanan, J., Friedrich, L., Jacklin, A., Chen, C., & Zickuhr, K. (2012). *How teens do research in the digital world.* Retrieved from http://www.pewinternet.org/2012/11/01/how-teens-do-research-in-the-digital-world/

Renzulli, J. S. (1976). The enrichment triad model: A guide for developing defensible programs for the gifted and talented. *Gifted Child Quarterly, 20,* 303–326.

Renzulli, J. S. (1978). What makes giftedness? Reexamining a definition. *Phi Delta Kappan, 60,* 180–184, 261.

Renzulli, J. S., Leppien, J. H., & Hays, T. S. (2000). *The multiple menu model: A practical guide for developing differentiated curriculum.* Waco, TX: Prufrock Press.

Renzulli, J. S. & Reis, S. M. (1997). *The Schoolwide Enrichment Model: A how-to guide for educational excellence* (2nd ed.). Waco, TX: Prufrock Press.

Reynolds, R., & Caperton, I. H. (2011). Contrasts in student engagement, meaning-making, dislikes, and challenges in a discovery-based program of game design learning. *Educational Technology Research and Development, 59,* 267–289.

Rockland, R., Bloom, D. S., Carpinelli, J., Burr-Alexander, L., Hirsch, L. S., & Kimmel, H. (2010). Advancing the "E" in K-12 STEM education. *Journal of Technology Studies, 36*(1), 53–64.

Sahin, A, Ayar, M. C., & Adiguzel, T. (2014). STEM related after-school program activities and associated outcomes on student learning. *Educational Sciences: Theory & Practice, 14*(1), 309–322.

Shea, D., Lubinski, D., & Benbow, C. P. (2001). Importance of assessing spatial ability in intellectually talented young adolescents: A 20-year longitudinal study. *Journal of Educational Psychology, 93,* 604–614.

Tomlinson, C. A., Kaplan, S. N., Renzulli, J. S., Purcell, J. H., Leppien, J. H., Burns, D. E., . . . Imbeau, M. B. (2009). *The parallel curriculum: A design to develop learner potential and challenge advanced learners* (2nd ed.). Thousand Oaks, CA: Corwin Press.

VanTassel-Baska, J. (2011). An introduction to the Integrated Curriculum Model. In J. VanTassel-Baska & C. A. Little (Eds.), *Content based curriculum for high-ability learners* (2nd ed., pp. 9–32). Waco, TX: Prufrock Press.

Verner, I. M. (2004). Robot manipulations: A synergy of visualization, computation, and action for spatial instruction. *International Journal of Computers for Mathematical Learning, 9,* 213–234.

Wai, J., Lubinski, D., & Benbow, C. P. (2009). Spatial ability for STEM domains: Aligning over 50 years of cumulative psychological knowledge solidifies its importance. *Journal of Educational Psychology, 101,* 817–835.

Webb, R. M., Lubinski, D., & Benbow, C. P. (2007). Spatial ability: A neglected dimension in talent searches for intellectually precocious youth. *Journal of Educational Psychology, 99,* 397–420.

Westberg, K. L., & Daoust, M. E. (2003, Fall). The results of the replication of the classroom practices survey replication in two states. *The National Research Center on the Gifted and Talented Newsletter,* 3–8. Retrieved from http://www.gifted.uconn.edu/nrcgt/newsletter/fall03/fall032.html

Whitton, D. (1997). Regular classroom practices with gifted students in grades 3 and 4 in New South Wales, Australia. *Gifted Education International, 12,* 34–38.

CHAPTER 5

E is for Engineering Education
Cultivating Applied Science Understandings and Problem-Solving Abilities

Debbie Dailey, Ed.D., & Alicia Cotabish, Ed.D.

Introduction

The Next Generation Science Standards (NGSS) promote student understanding of science content and concepts through the practices of scientific inquiry and engineering design. Students are expected to demonstrate their understanding of science by investigating the natural world and developing solutions to meaningful problems. Through these actions, students should be given ample opportunities to practice creative and critical thinking skills as they develop into real-world investigators. As students progress in grade levels, the scientific and engineering practices increase in complexity and sophistication. For example, in grades K–2, students begin defining problems and testing and comparing possible solutions to simple problems. By the time they enter high school, students will investigate problems of social and global significance and use quantitative methods and/or computer stimulations to compare and test possible solutions. Advanced learners, including those from diverse backgrounds, should thrive in opportunities to investigate relevant problems of social and global significance.

 DOI: 10.4324/9781003238218-6

Why Engineering

The technological advancement in our world over the past 20+ years is astonishing. In past years, cellular phones and computers were considered a luxury in most households. Today, cellular phones serve as personal computers and most households have access to some type of computer device. With the advancing technologies, various types of engineering jobs and careers are abundant, and there is a lack of employees to meet the increasing demands (Change the Equation, 2012). As Milano (2013) pointed out, "Students entering Kindergarten this year will likely enter job fields upon graduation that have not yet been developed, using knowledge that has not yet been discovered and tools that have not yet be engineered" (p. 11). With this in mind, it is the responsibility of our educational system to prepare learners for the advancements in technology and engineering.

. . . by the end of 12th grade, " . . . students should have gained sufficient knowledge of the practices, crosscutting concepts, and core ideas of science and engineering to engage in public discussions on science-related issues, to be critical consumers of scientific information related to their everyday lives, and to continue to learn about science throughout their lives" (NRC, 2012, p. 12).

The authors of the *A Framework for K–12 Science Education*, the foundation for the Next Generation Science Standards, placed engineering on equal footing as scientific inquiry by integrating engineering into the practices and disciplinary core ideas (National Research Council [NRC], 2012). They believed in the importance of students discovering and experiencing the practical uses of science through engineering and technology, and for this reason, the practices of engineering were combined with the practices of science. The goal of the *Framework,* and hence the NGSS, is not to necessarily develop more engineers, but by the end of 12th grade, " . . . students should have gained sufficient knowledge of the practices, crosscutting concepts, and core ideas of science and engineering to engage in public discussions on science-related issues, to be critical consumers of scientific information related to their everyday lives, and to continue to learn about science throughout their lives" (NRC, 2012, p. 12). By providing students with experiences in both science and engineering practices, students will be more prepared to solve societal and environmental problems that they encounter throughout their lives.

Scientists Versus Engineers

Naturally, scientists and engineers complement each other's work. Scientists typically make testable predictions and explanations about the world, and engineers identify a specific need and design a solution to meet that need. For example, a scientist may observe, gather information, hypothesize, and test for a treatment or cure for cancer, whereas the engineer may design the equipment needed for the test or develop the drug used to treat the cancer. Often, engineering is referred to as *applied science*. In other words, engineers apply the knowledge of science to address the need and design a solution.

> Often, engineering is referred to as *applied science*. In other words, engineers apply the knowledge of science to address the need and design a solution.

Figures 5.1 and 5.2 compare a typical scientific method and engineering method. It is important to note that neither method should be considered a linear process—rather, each step in the method can and should be revisited multiple times. As you can see, both methods are similar and require evidence resulting from the analysis of collected data.

Engineering and the NGSS

The authors of *A Framework for K–12 Science Education* proposed eight practices (see Table 5.1) to be included in the NGSS that describe behaviors that scientists and engineers engage in as they investigate the natural world or design and build models and systems to address a need (Achieve, Inc., 2014). These practices should not be taught in isolation but integrated with content (disciplinary core ideas) and crosscutting concepts to explain a particular phenomenon.

The practices are intended for all students in grades K–12, but they increase in complexity and sophistication across grade levels (see Figure 5.3). For example, grades K–2 students are asked to define a simple problem that can be solved through the development of a new tool or refinement of an existing tool. Students in grades 3–5 are instructed to use prior knowledge to identify an existing problem that can be solved through the development of a new tool. Students in grades 6–8 are further challenged by the addition of multiple criteria and constraints limiting the possible solutions. For students in grades 9–12, the complexity and sophistication increase by the addition of social, techni-

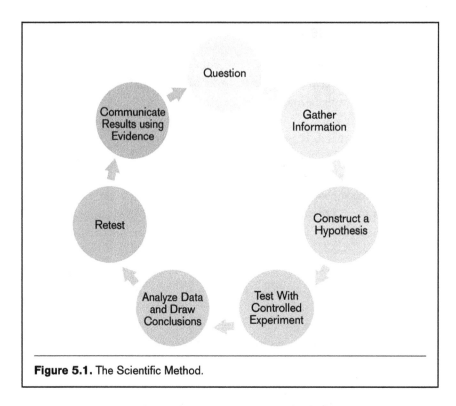

Figure 5.1. The Scientific Method.

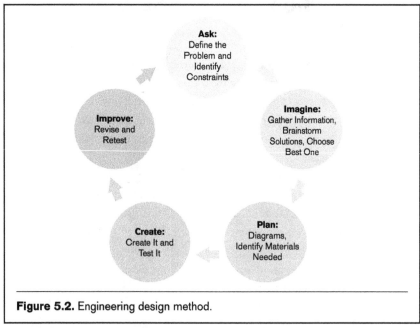

Figure 5.2. Engineering design method.

TABLE 5.1
Science and Engineering Practices

Asking Questions (for science) and Defining Problems (for engineering)
Developing and Using Models
Planning and Carrying Out Investigations
Analyzing and Interpreting data
Using Mathematics and Computational Thinking
Constructing Explanations (for science) and Designing Solutions (for engineering)
Engaging in Argument from Evidence
Obtaining, Evaluating, and Communicating Information

From *A Framework for K–12 Science Education: Practices, Crosscutting Concepts, and Core Ideas* (p. 3), by National Research Council, 2012, Washington, DC: The National Academies Press.

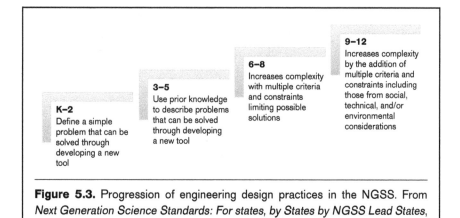

Figure 5.3. Progression of engineering design practices in the NGSS. From *Next Generation Science Standards: For states, by States by NGSS Lead States*, 2013, Washington, DC: The National Academies Press.

cal, and/or environmental considerations (NGSS Lead States, 2013).

The progression and development of the science and engineering practices across grade levels are referred to as learning progressions (NGSS Lead States, 2013). NGSS also described learning progressions for the two other dimensions: disciplinary core ideas and crosscutting concepts. The learning progressions provide a general trajectory for typical learners but can be differentiated as needed for gifted and advanced science learners (Adams, Cotabish, & Dailey, 2015). Furthermore, the learning

For example, a student in grades K–2 that demonstrates an ability to solve a simple problem can be challenged by adding constraints or specific criteria to the problem or being asked to use prior knowledge to identify a problem that needs solving.

progressions can be used to accelerate or add complexity as students advance—regardless of grade level. For example, a student in grades K–2 that demonstrates an ability to solve a simple problem can be challenged by adding constraints or specific criteria to the problem or being asked to use prior knowledge to identify a problem that needs solving.

Integrating Engineering into Existing Curriculum

Integrating engineering practices into existing curriculum is doable and efficient. The good news is that many teachers in gifted and talented classrooms already integrate activities that align with engineering practices into their existing curriculum. For example, it is common to find teachers of the gifted challenging advanced learners with protecting an egg in an egg-drop challenge, designing and building a sturdy bridge with straws, and creating the tallest tower with toothpicks. Unfortunately, these activities are often just activities and are rarely embedded in science content. To make these activities more meaningful and relevant to students, they should be tied to science content and be related to a real-world problem. Project and problem-based learning are excellent avenues for integrating engineering into science content combined with a real-world problem or issue.

Project and Problem-Based Learning

According to VanTassel-Baska and the Educational Resources Information Center (1998), science curriculum for gifted learners should emphasize inquiry, especially problem-based and project-based learning (PBL). The more that students can construct their understanding about science for themselves, the better able they will be to encounter new situations and apply appropriate scientific processes to them. Problem- and project-based learning can serve as a catalyst to create these types of encounters. Furthermore, the integration of technology as a learning tool is prevalent among the two PBLs. The use of technology to teach science offers some exciting possibilities for connecting students to real-world opportunities. From a scientific process and experimental design view, problem- and project-based learning create an avenue to explore scientific processes, using experimental design procedures. In particular, both PBLs would require students to investigate a particular topic, test a hypothesis, and then follow through with appropriate procedures, participate in discussions, engage in the re-analysis of the problem, and communicate their findings to a relevant audience. Unlike curricular add-ons at the end, problem- and project-based learning is the context for learning.

So what are the differences between the two PBLs? Finkle and Torp (1995) described problem-based learning as a "… curriculum development and instructional system that simultaneously develops both problem solving strategies and disciplinary knowledge bases and skills by placing students in the active role of problem solvers confronted with ill-structured problem that mirror real-world problems" (p. 1). The focus of problem-based learning is solving a real-world problem, whereas students involved in project-based learning will work together to design and possibly create a product to address a real-world need. Figure 5.4 presents the similarities and differences between the two PBLs.

In problem-based learning, students are presented real-world problems through some type of scenario. To increase the relevancy of the learning, the problem may be tied to an authentic community-based issue. Once the scenario is presented, students identify the significant problem and work to derive a solution through a process of six steps (Finkle & Torp, 1995). Adams, Cotabish, and Dailey (2015) proposed a seventh (optional) step, resolution/action, because an additional step may be necessary to carry out the solution. The seven steps are:

1. Introduce an ill-structured problem.
2. Identify the three "What's," also known as "Need to Knows" or "KWL Chart." Students will list what is known, what they need to know, and what they need to do.
3. Gather information.
4. Hypothesize possible solutions.
5. Determine the best-fit solution.
6. Present the solution.
7. Resolution/action (optional).

Steps 2–5 are not necessarily sequential and may be conducted simultaneously, as new information may redefine the original problem. Near the conclusion of the PBL experience, the class comes to an agreement on the best-fit solution. Ideally, the resolution of a community-based problem will lead to action, such as a service-learning project.

In project-based learning, the teacher begins the instruction with essential/driving questions to direct students through the investigation. The teacher scaf-

Problem-based Learning versus Project-based Learning	
Similarities	
Both: Provide opportunities for differentiation Are open-ended in nature Address 21ˢᵗ Century Learning competencies Are task driven Employ entry events Typically conducted in groups Are student-centered Used as a formative assessment Includes the 3 "What's" or "Need to Knows" Involve research of subject matter Spur in-depth inquiry Follows steps Prompt critical and creative thinking	
Differences	
Problem-based Learning	**Project-based Learning**
Typically shorter in duration	Often longer in duration
Choice is tied to possible solutions	Frequently employs student choice throughout
Often single subject	Often interdisciplinary / integrative
Products are often in the form of solutions	Emphasis on final product
Multiple paths for solving ill-structured problem	Centered around driving questions
New-found information may redirect or pose additional questions	Final products are often presented to public audiences
Often uses case studies or fictitious scenarios to set up the problem	Typically involves real-world problem
May require an additional action step to carry out and resolve the issue(s)	Employs revision and reflection
May or may not utilize technology	Utilizes technology

Figure 5.4. Problem-based Versus Project-based Learning. From from *A Teacher's Guide to Using the Next Generation Science Standards with Gifted and Advanced Learners in Science* (p. 99), by C. Adams, A. Cotabish, & D. Dailey, 2015, Waco, TX: Prufrock Press. Copyright 2015 by National Association of Gifted Children. Reprinted with permission.

folds the learning for students through labs, lectures, technology applications, and instructional activities. Throughout the investigation, students create and continuously revisit a "Need to Know" list. As the investigation concludes, students reflect and consider peer and teacher feedback to better inform their learning. As a final activity, students present their products or creations to an audience—ideally, to professionals in a related field of study.

As students progress through either PBL process, they develop 21st-century knowledge, skills, and dispositions in self-directed learning, critical thinking, problem solving, innovative thinking, and collaboration. Both PBLs lend themselves to meeting the eight Science and Engineering Practices from the NGSS/NRC framework for K–12 science education, if planned accordingly. For example, to meet the NGSS Standard K–2-ETS1-1 (i.e., ask questions, make observations, and gather information about a situation people want to change to define a simple problem that can be solved through the development of a new or improved object or tool), students could engage in a PBL that would require them to respond to the following essential question: *How do plants and/or animals use their external parts to help them survive, grow, and meet their needs?*

Students could demonstrate (through a design challenge) their understanding of the function of plant and animal external parts by using them to solve human problems. Although this particular challenge could work with either PBL, the sequential process would be slightly different for each PBL, and the problem may or may not be grounded in a fictitious scenario. As a project-based learning activity, the culminating product would require a presentation to an appropriate audience (e.g., professional in the field). Regardless of which PBL format is utilized, both can serve as a valuable framework for engaging students in engineering practices.

Using the Standards to Integrate Engineering Into the Curriculum

Using the NGSS standards to integrate engineering into the curriculum requires planning. Milano (2013) provided a great example of how to bundle NGSS performance expectations with instructional questions to guide student learning (see Table 5.2). Situated in a real-world problem, students are learning about weather and heating effects of the sun while engaging in authentic science and engineering practices.

TABLE 5.2
Weather, Climate, and Engineering Design Bundle

Weather, Climate, and Engineering Design Bundle		Instructional Questions
K-PS3-1	Make observations to determine the effect of sunlight on Earth's surface. [Clarification Statement: Examples of Earth's surface could include sand, soil, rocks, and water] [*Assessment Boundary: Assessment of temperature is limited to relative measures such as warmer/cooler.*]	What happens to the sand in the sandbox when the sun shines on it all morning?
K-2-ETS1-1	Ask questions, make observations, and gather information about a situation people want to change to define a simple problem that can be solved through the development of a new or improved object or tool.	Students cannot sit in the sandbox because the sand is too warm. What do we need to know to help solve this problem?
K-PS3-2	Use tools and materials to design and build a structure that will reduce the warming effect of sunlight on an area.* [Clarification Statement: Examples of structures could include umbrellas, canopies, and tents that minimize the warming effect of the sun.]	What type of structure will keep the sand in the sandbox the coolest?

Adapted from "The Next Generation Science Standards and Engineering for Young Learners: Beyond Bridges and Egg Drops," by M. Milano, 2013, *Science and Children, 51*(2), 10–16. Copyright 2013 by National Science Teachers Association. Adapted with permission.

Notice the variation in the instructional questions. In the first question, students are asked to predict what will happen to the temperature of the sand as the sun shines on it all day. In regard to the second question, students are asked to find information to resolve the hot sand problem. In the third question, students are asked to consider a structure that would help them solve the problem. This example demonstrates how to integrate the practices within the content.

Using Milano's table, we have bundled another set of performance expectations and developed the corresponding instructional questions (Table 5.3). Again, the instructional questions are situated in a real problem in the context of energy and energy transfer. Students are asked to engage in both science and engineering practices as they work to develop a device that will conserve energy.

Conclusion

Engineering practices can offer a rich context for developing many 21st-century skills, such as critical thinking, problem solving, and information

TABLE 5.3

Energy, Energy Transfer, and Engineering Design Bundle

Energy, Energy Transfer, and Engineering Design Bundle		Instructional Questions
MS-PS3-3.	**Apply scientific principles to design, construct, and test a device that either minimizes or maximizes thermal energy transfer.*** [Clarification Statement: Examples of devices could include an insulated box, a solar cooker, and a Styrofoam cup.] [*Assessment Boundary: Assessment does not include calculating the total amount of thermal energy transferred.*]	What kind of device would keep your cup of coffee warm during a cold football game?
MS-PS3-4.	**Plan an investigation to determine the relationships among the energy transferred, the type of matter, the mass, and the change in the average kinetic energy of the particles as measured by the temperature of the sample.** [Clarification Statement: Examples of experiments could include comparing final water temperatures after different masses of ice melted in the same volume of water with the same initial temperature, the temperature change of samples of different materials with the same mass as they cool or heat in the environment, or the same material with different masses when a specific amount of energy is added.] [*Assessment Boundary: Assessment does not include calculating the total amount of thermal energy transferred.*]	How does the kinetic energy of your coffee compare in the devices that you designed and created?
MS-PS3-5	**Construct, use, and present arguments to support the claim that when the kinetic energy of an object changes, energy is transferred to or from the object.** [Clarification Statement: Examples of empirical evidence used in arguments could include an inventory or other representation of the energy before and after the transfer in the form of temperature changes or motion of object.] [*Assessment Boundary: Assessment does not include calculations of energy.*]	What happened to the kinetic energy of the molecules in the coffee when the coffee was allowed to cool? What evidence do you have to support your claim? (Where did the heat energy go?)
MS-ETS1-4	Develop a model to generate data for iterative testing and modification of a proposed object, tool, or process such that an optimal design can be achieved.	How can your device be improved?

Adapted from "The Next Generation Science Standards and Engineering for Young Learners: Beyond Bridges and Egg Drops," by M. Milano, 2013, *Science and Children, 51*(2), 10–16. Copyright 2013 by National Science Teachers Association. Adapted with permission.

literacy, especially when instruction addresses the nature of science and promotes use of engineering practices. Cultivating applied science understandings and problem-solving abilities are important considerations for curriculum planning. Although the integration of engineering practices are new to the NGSS, the concepts of content integration, differentiation, and facilitating applied understanding through problem-solving activities are not new concepts to the field of gifted education—in fact, they should be second nature! The biggest challenge to gifted educators in regard to integrating engineering practices into existing curriculum will be to embed the applied activities into science content. Although we recognize that science may or may not be an area of expertise for all gifted educators, it is reassuring to know that engineering is an applied discipline that serves students well when student directed.

Discussion Questions

1. Considering the performance expectations and instructional questions in Table 5.3, how could you differentiate the content, process, or product for gifted or advanced science learners?
2. In regard to integrating engineering practices, how could you transform your existing PBLs into science-focused engineering activities?

References

Achieve, Inc. (2014). *Next Generation Science Standards.* Washington, DC: Author.

Adams, A., Cotabish, A., & Dailey, D. (2015). *A teacher's guide to using the Next Generation Science Standards with gifted and advanced learners.* Waco, TX: Prufrock Press.

Change the Equation. (2012). *Vital signs: Reports on the condition of STEM learning in the U.S.* Retrieved from http://changetheequation.org/sites/default/files/CTEq_VitalSigns_Supply(2).pdf

Finkle, S. L., & Torp, L. L. (1995). *Introductory documents.* Aurora, IL: Center for Problem-Based Learning.

Milano, M. (2013). The Next Generation Science Standards and engineering for young learners: Beyond bridges and egg drops. *Science and Children, 51*(2), 10–16.

National Research Council. (2012). *A framework for K–12 science education: Practices, crosscutting concepts, and core ideas.* Washington, DC: The National Academies Press.

NGSS Lead States. (2013). *Next Generation Science Standards: For states, by states.* Washington, DC: The National Academies Press.

VanTassel-Baska, J., & Educational Resources Information Center. (1998). *Planning science programs for high ability learners.* Reston, VA: ERIC Clearinghouse on Disabilities and Gifted Education, Council for Exceptional Children.

CHAPTER 6

M is for Mathematics at the Elementary Level

Scott A. Chamberlin, Ph.D., & Eric L. Mann, Ph.D.

It may be the case that for many years elementary mathematics teachers were blessed with the leisure of preparing learners to acquire knowledge in mathematics with little recompense for how the mathematics would actually be used. Given the recent emphasis on STEM (Robelen, 2010), mathematics learning and instruction has assumed a place of prominence. It could be argued, therefore, that greater accountability may be requisite to ensure that learners can utilize mathematics concepts to solve problems in science, technology, engineering, and mathematics, rather than just mathematics. Several factors figure into the STEM learning equation. Many decades ago, it was agreed by educational psychologists that myriad factors affect learning (Good & Brophy, 1986). Certainly, some of the factors are teacher effectiveness and approach, learner characteristics and background, and influences such as parental educational background. Arguably, one of the most salient characteristics that affects the learning equation is the selected curriculum. All other factors being considered equal, the better the curriculum, the greater the learning outcomes should be.

This chapter is comprised of two sections. First, desirable characteristics of model curricula in mathematics education are discussed. Subsequently, four curricula that meet such characteristics are discussed. The chapter concludes

 DOI: 10.4324/9781003238218-7

with a recommendation of the *ideal* mathematics curriculum for STEM development of gifted learners.

Elementary Mathematics Curriculum Designed to Develop Talent in a Sequenced STEM Plan

Although the topic of requisite curricular characteristics in mathematics that engenders meaningful learning in STEM disciplines is yet to be ascertained empirically in one study, much attention has been invested in components of mathematics curricula. For instance, the notion that conceptual learning in mathematics is a necessity is generally agreed upon by experts in the field of mathematics, and it has been shown empirically to be effective in promoting understanding of mathematics (Richland, Stigler, & Holyoak, 2012).

Conceptual Learning

If conceptual learning is the main objective in mathematics learning and the explicit curriculum largely determines what is taught in the classroom, it makes sense that mathematics curriculum must be conceptually based. In fact, the development of conceptual learning in mathematics is considered integral to the development of flexible learning (Krutetskii, 1976). Such characteristics are often considered imperative to understanding higher level mathematics and to the production of creative outcomes because without such qualities, young learners may have greatly compromised abilities in solving problems. Given the fact that problem solving is considered the very essence of mathematics (National Council of Supervisors of Mathematics [NCSM], 1977; National Council of Teachers of Mathematics [NCTM], 1980), the lack of conceptual understanding and flexibility in thinking may hinder young learners' progress. Curricula that rely heavily on teachers and learners memorizing procedures have therefore fallen out of favor among many mathematics educators. Consequently, the question among mathematics educators of the gifted often remains, "So, what do I have to do to precipitate conceptual learning in mathematics or does this just happen automatically?" Certainly, conceptual learning in mathematics does not happen automatically. Teachers and learning facilitators need to be deliberate in their chosen approaches and curricula.

Mathematical Problem Solving

One characteristic that must be sought in any curriculum is a high degree of problem solving (Hiebert et al., 2000). When mathematics educators specifically seek problem-solving tasks, they prepare mathematics and science learners for responsibilities that would ordinarily be incurred when doing mathematics (Lesh et al., 2000). With Model-Eliciting Activities (MEAs), Lesh and colleagues advertised that aspiring mathematicians, such as gifted and talented mathematicians, engage in precollege level thinking in mathematics and science. It is a safe assumption that to succeed in a world filled with complex problems, learners cannot be expected to simply memorize algorithms and procedures in mathematics and be able to apply them magically. Learners thus need to engage in legitimate problem solving tasks so that they can create their own understanding of mathematics (for a comprehensive overview of how mathematics educators in the western hemisphere define "problem solving," see Chamberlin, 2008). The aforementioned components of mathematics curricula for advanced students should facilitate conceptual learning and flexibility in thinking through mathematical problem solving and are components that serve students well in several capacities. It has already been shown that the use of mathematical problem-solving tasks with realistic contexts may precipitate a high degree of interest among advanced mathematicians (Chamberlin, 2002).

> . . . components of mathematics curricula for advanced students should facilitate conceptual learning and flexibility in thinking through mathematical problem solving and are components that serve students well in several capacities.

Multiple Entry Points

Several criteria should be solicited when identifying what types of mathematical problem-solving tasks to use with aspiring STEM students in elementary grades; many of these were outlined by Chamberlin (2008). One component that deserves attention is that to serve varied constituencies (i.e., young gifted STEM students in an inclusion, pullout, or magnet setting), problem-solving tasks should have multiple entry points. The term *multiple entry points* is confusing to many in the world of education. Simply stated, multiple entry points in mathematics suggest that students of varying abilities may be served well with problems that may be solved through various levels of mathematics. As an example, a problem that pertains to Historic Hotels, can be found at the following link: http://www.region11mathandscience.org/archives/files/Problem%20Solving/Teachers/Historic%20HotelsMEAtchrmaterials.pdf. It is one in which students are asked to compute minimum/maximum values. The problem statement for Historic Hotels is included below.

> Frank Graham, from Elkhart, IN, has just inherited a historic hotel. He would like to keep the hotel, but he has little experience in hotel management. The hotel has 80 rooms, and Mr. Graham was told by the previous owner that all of the rooms are occupied when the daily rate is $120 per room. He was also told that for every dollar increase in the daily $120 rate, one fewer room is rented. So, for example, if he charged $121 dollars per room, only 79 rooms would be occupied. Each occupied room has a $4 cost for service and maintenance per day in addition to the daily fee.
>
> Frank would like to know how much he should charge per room in order to maximize his profit and what his profit would be at that rate. In addition, he would like to have a procedure for finding the daily rate that would maximize his profit in the future even if the hotel prices and the maintenance costs change. Write a letter to Mr. Graham telling him what price to charge for the rooms to maximize his profit and include your procedure for him to use in the future.

This business-related problem can be solved with several solutions and has been solved by students as young as grade 6 using number sense and operations. Similarly, it has been solved using somewhat sophisticated procedures in calculus by graduate students in mathematics education (personal communication with R. Lesh, 2014). This task is one that has multiple entry points because problem solvers with considerably different levels of knowledge in the field of mathematics can solve the same problem using less or more complex processes. To a teacher that desires to serve multiple constituents, perhaps including general population students and gifted students simultaneously, problem-solving tasks that have multiple entry points can be invaluable.

Creativity and Ingenuity

One attribute of aspiring mathematicians that may be neglected often, though, is the construct of creativity and ingenuity. This construct is instrumental in the development of new products, creations, and inventions. Without adequate means to create, the world of STEM is at an impasse. Myriad individuals in the world of STEM have reflected that they do not need human calculators; they need humans capable of thinking. Köhler (1997) stated that, "Modern day commerce has no use for pupils graduating from school who have been trimmed to mechanically solve problems in exactly one pre-given way, i.e., like a mechanic" (p. 89). Further, there is converging evidence that students often emerge from K–12 mathematics education with adequate problem execution skills—but with inadequate problem representation skills (Mayer & Hegarty, 1996). Simply stated, humans cannot out-process computers and software. However, they can identify needs in the STEM industry and start working on means by which to meet such needs. This is accomplished through ingenuity and creativity. Individuals who can execute problem-solving skills, as Mayer and Hegarty stated (1996), and, in a mechanical fashion, as Köhler (1997) stated, would have little chance of success in applying mathematics to STEM problems at this level of complexity because the answers are not readily apparent. Hence, a curriculum rich in mathematical problem-solving experiences is requisite for advanced learners for many reasons. To reiterate, if learners are expected to think like STEM experts in their vocation, they need to be expected to think like them (i.e., creatively) during learning episodes.

> One attribute of aspiring mathematicians that may be neglected often, though, is the construct of creativity and ingenuity . . . . Myriad individuals in the world of STEM have reflected that they do not need human calculators; they need humans capable of thinking . . . . Simply stated, humans cannot out-process computers and software. However, they can identify needs in the STEM industry and start working on means by which to meet such needs. This is accomplished through ingenuity and creativity . . . . if learners are expected to think like STEM experts in their vocation, they need to be expected to think like them (i.e., creatively) during learning episodes.

Critical Thinking

Several additional characteristics are endemic in *best* mathematics curricula. For instance, exemplary mathematics curricula generally incorporate some consideration of critical thinking (Kettler, 2014). Critical thinking, according to Facione (1990), is comprised of such regulatory skills as interpretation, analysis, evaluation, inference, and explanation. Finding curricula that genuinely support and facilitate critical thinking skills would appear to be a simplistic accomplishment given the recent emphasis on it. This is evidenced by the fact that the past three mathematics standards documents have, in some form or another, called for critical thinking skills to be engendered in the classroom (NCTM, 1989; 2000; NGA & CCSO, 2010). However, what is advertised in curricula and what is actually presented in curricula does not always align. It could be argued that the single best method of facilitating critical thinking in mathematics is to have learners solve problems in their own manner and then identify which solutions are the most sophisticated and elegant (Boaler, 2002). Krutetskii (1976) referred to this process as appreciating the aesthetics of mathematics.

Mathematical Modeling

Proponents of mathematical modeling have suggested its significance since the mid-1970s (Lesh & Johnson, 1976). In 2010, the authors of the Common Core State Standards in mathematics made it an expectation as one of the eight mathematical practices. It appears as though policy makers in mathematics education have recognized the value of mathematical modeling. Mathematical modeling enjoys many conceptions. Perhaps the most simplistic analogy to mathematical modeling is the development of formulae or algorithms in mathematics. Although this analogy is not a perfect one (i.e., mathematical modeling is not merely the production of formulae or algorithms), the activities do hold parallels in that each process leaves residue among problem solvers because they have created a representation of mathematics that can be used in subsequent, similar settings. Lesh and colleagues (2000) referred to mathematical modeling as a system of elements, relationships (among elements), operations, and patterns or rules. Ideally, these systems explain or describe another system.

Mathematical modeling activities such as Model-Eliciting Activities (MEAs) are a subset of problem-solving tasks that have a precise set of six principles for design (Lesh et al., 2000). They are highly open-ended and leave opportunity for multiple solutions. Learners generally solve MEAs in groups of three and they are expected to communicate their solutions to peers. Communicating solutions to peers places a high demand on students to have a mathematical model that may not be simplistic, but that is easy to interpret (NGA & CCSO, 2010). Lesh and colleagues have referred to the process of mathematical modeling as capturing some of the most powerful ideas in precollege level mathematics and science (which subsumes STEM). It is for this reason that MEAs are a strong match for developing STEM aspirants.

Acceleration and Enrichment

The debate about whether acceleration (Colangelo, Assouline, & Gross, 2004) or enrichment (Rickansrud, 2011) best serves gifted STEM learners is yet to be resolved. This is likely the case because learners possess such individual characteristics that a gross overgeneralization, such as only acceleration or enrichment will serve all students best in every case, is avoided by academicians. In short, enrichment possesses some characteristics that help students maximize development and in other instances, acceleration or advancement is the best option. One notion on which most curricular experts agree is that learning episodes in which acceleration and enrichment can occur are ideal. It is for this reason that a few select curricular approaches, such as MEAs, accommodate learners that require acceleration and enrichment.

> . . . curricular experts agree . . . that learning episodes in which acceleration and enrichment can occur are ideal. It is for this reason that a few select curricular approaches, such as MEAs, accommodate learners that require acceleration and enrichment.

The Future of Curricula to Develop Mathematical Talent in STEM Disciplines

The question remains, "What constitutes the ideal curriculum for mathematical development in STEM disciplines?" It could be argued that no such ideal curriculum exists, but others might argue that some common characteristics exist that should be present in most curricula. As an example, curricula with abundant open-ended problem solving activities that foster enrichment, acceleration, mathematical modeling, critical and creative thinking, conceptual learning, and that have multiple entry points would be ideal. The term *ideal* connotes something that may not be attainable. Consequently, teachers of elementary mathematics students in STEM disciplines should seek and expect as many of these characteristics as possible. To further obfuscate the issue, the ideal curriculum would meet as many state and national standards (e.g., NCTM, 2000) as possible, including the Common Core State Standards in mathematics. Two components of the existing national standards, content and process, should be heeded to be a comprehensive one.

In 1989 and again in 2000, authors of the *Principles and Standards for School Mathematics* (NCTM, 2000), divided mathematics into five content areas: number (sense) and operations, geometry, measurement, algebra, and data analysis (statistics) and probability. Authors of the Common Core State Standards in mathematics (CCSS-M) divided mathematics into 11 domains, including: counting and cardinality, operations and algebraic thinking, number and operations in base ten, number and operations-fractions, measurement and data, geometry, ratios and proportional relationships, the number systems, expressions and equations, functions, and statistics and probability. As much as possible, the ideal curriculum should take into consideration the five content areas and 11 domains, though much overlap occurs in the two conceptions of mathematics as content areas.

> . . . curricula with abundant open-ended problem solving activities that foster enrichment, acceleration, mathematical modeling, critical and creative thinking, conceptual learning, and that have multiple entry points would be ideal.

Similarly, the *Principles and Standards for School Mathematics* contain five process standards (what learners should be doing) and they include the following: communication, connections, problem solving, reasoning and proof, and representation. The CCSS-M has altered the five process standards to reflect eight mathematical practices. These practices consist of the following: Make sense of problems and persevere in solving them, reason abstractly and quantitatively, construct viable arguments and critique the reasoning of others, model with mathematics, use appropriate tools strategically, attend to precision, look for and make use of structure, look for and express regularity in repeated reasoning. Comprehensive curricula should take into consideration the content and process standards and expectations, with the realization that many of the aforementioned standards are duplicated in the two standards documents.

Four Sample Curricula

The four sample curricula discussed in the following section range from supplementary curricula to a fully operational, stand-alone curriculum. Each is discussed with brief commentary on its advantages and disadvantages. As curricula were analyzed for this chapter, the curriculum components discussed earlier were used as the metric for inclusion.

Building Blocks©

The first curriculum discussed is one created by Doug Clements and Julie Sarama (2013) entitled Building Blocks©. Development of the curriculum is predicated on the most recent research in learning, called learning trajectories. It is the outcome of a National Science Foundation grant, and the electronic media was created to individually serve students. Interestingly, much like dynamic assessments, Building Blocks© adjusts activities based on the last learner response. That is to say, if a learner has a misconception or exhibits proficiency, the software responds accordingly.

The premise behind the software and the printed curriculum is to isolate learners' areas of need and to help refine those areas so that proficiency is achieved. Assessment data is provided to teachers so that they can guide learners. A redeeming quality of the software is that for individual use, it is extremely affordable. Hard copies of the curriculum also exist and with it, the onus is on the instructor to differentiate. It is for this reason that Building Blocks© holds promise for the development of mathematical knowledge and can easily be used in home. Teacher and district use can also be purchased. In addition, the software appears to be highly interactive and incorporates liberal use of graphics to simultaneously entertain and challenge young learners in elementary grades. Relative to content areas, Building Blocks© contains activities in all five NCTM (2000) content areas and it is aligned with CCSS-M standards. It also emphasizes conceptual thinking and differentiation through problem solving. Few mathematics curricula that are research based exist to serve students in early grades so this curriculum is very exciting and fills a niche. Presently, the grade 1 curriculum is near completion and plans are being developed for mathematics learners in grades 2–5. To learn about Building Blocks©, visit https://www.mheonline.com/programMHID/view/0076BB2012.

Model-Eliciting Activities

Model-Eliciting Activities (MEAs) were briefly mentioned earlier in the chapter. Like Building Blocks©, MEAs were not developed with the intent to serve as a stand-alone curriculum. MEAs were intended to supplement curricula by challenging students to create mathematical models. The intent of designing mathematical models is to help learners make sense of mathematics. MEAs possess great potential for facilitating and identifying creativity (Coxbill, Chamberlin, & Weatherford, 2013). Although MEAs were not initially designed for use with learners gifted in mathematics, significant efforts have been invested by several researchers to ensure that they are crafted with gifted students in mind. MEAs are open-ended problem-solving tasks that are designed using six principles:

  ▷ The model construction principle: The model created must describe, explain, and/or justify a prediction;
  ▷ The reality principle: The model must create meaningfulness by enabling learners to make sense of the mathematical phenomena being investigated;
  ▷ The self-assessment principle: The model created must be able to inform learners in the group of the usefulness of their solution;
  ▷ The construct documentation principle: The model created must help teachers interpret the manner in which problem solvers are responding through a documented trail;
  ▷ The construct shareability and reusability principle: The model created must be one that shows use in various similar situations and is therefore generalizable; and
  ▷ The effective prototype principle: The model must be one that is simplistic enough to answer the question posed (in the problem), but complex enough to be used in additional mathematics settings.

With respect to development of MEAs and research, Chamberlin (2002, 2008, 2013, in press) has likely invested the most amount of time on MEAs with gifted students. In fact, he created learning activities relevant to statistics and is creating activities relevant to probability as well. These activities have been developed with gifted students in mind and have been used mostly with upper elementary and middle-grade mathematics students. Moreover, his publications include teacher tips, including assessment discussions, designed to help MEA beginners. Probability and statistics have innumerable applications to STEM, including mathematics, physics, chemistry, engineering, and many other disciplines. A positive component to MEAs is that many of them can be found online. A caveat with some of the MEAs found online is that they may not have been developed with GT students in mind. One of the most robust databases of MEAs is housed at the University of Minnesota (see http://cehd-vision2020.umn.edu/cehd-blog/stem-meas-in-action/). Purdue University also

has a healthy selection of MEAs available (see https://engineering.purdue.edu/
ENE/Research/SGMM/CASESTUDIESKIDSWEB/case_studies_table.
htm). In addition, Chamberlin's book publications, *Statistics for Kids: Model-
Eliciting Activities to Investigate Concepts in Statistics* (2013) and the forthcom-
ing *Using MEAs to Investigate Concepts in Probability* (in press) provide teacher
tips for implementation, assessment hints and prospective solutions, links to
standards documents, and full-scale Model-Eliciting Activities specifically
developed with gifted learners as the priority. Chamberlin's publications can be
found at http://www.prufrock.com/.

Each MEA is comprised of four parts: a newspaper article, a set of readiness
questions, a problem statement, and a piece of mathematical information used
to solve the problem (such as a data table, map, etc.). After reading the newspaper article, learners respond to the questions. Oftentimes, the intent of the
questions is to define fuzzy constructs presented in the problem. Subsequently,
learners invest between 45 minutes to one hour solving the problem or creating
their mathematical model. A sample problem statement, not the full MEA, and
the accompanying data table are provided in Figure 6.1.

Project M² and M³

These stand-alone curricula are comprised of two separate, yet interrelated,
curricula. Project M^2 is an acronym for Mentoring Young Mathematicians and
Project M^3 is an acronym for Mentoring Mathematical Minds. As with all other
cited curricula in this chapter, each of these is a research-based and proven
approach with gifted and talented learners. The philosophy behind the respec-
tive curricula is to aggressively challenge young learners so that they can achieve
at high levels and thus be accelerated and enriched for later years. As with
Building Blocks© and MEAs, Projects M^2 and M^3 have differentiation built into
their curricula. In addition, the intent of each of the curricula is to expect young
learners to mimic what mathematicians actually do in the real world (similar to
the philosophical approach behind MEAs). In short, young mathematicians are
expected to make sense of mathematics (Hiebert et al., 2000).

Project M^2 is geometry and measurement related and has gifted grade
K–2 students respond to mathematical questions that are open-ended and
problem-solving based (Gavin, Casa, Adelson, & Firmender, 2013). Project
M^2 incorporated best practices from mathematics education, gifted education,
and early childhood education (Gavin, Casa, Firmender, & Carroll, 2013). The
development of this curriculum was predicated on the notion that young chil-

Problem Statement

Using the data provided about Coach Hall's athletes, create a method for awarding scholarships to future athletes interested in joining the team. It is reasonable to award two to three scholarships per year and it is possible to distribute less than a full scholarship. Remember, if you give too small of a scholarship, you may lose the athlete to another team, so you may want to consider awarding several different levels of scholarships (e.g., 100%, 75%, 50%, 25%). Your task is to provide a rationale for awarding the scholarships. Write a detailed letter to Coach Hall showing and explaining how you decided your rationale.

Data table

Athlete	High School 5,000 meter PR	High School 10,000 meter PR	College freshman PR	College sophomore PR	College junior PR	College senior PR	All-time PR
William Kiprono	14:32	30:06	30:38	29:27	29:36	30:14	29:27
Sean McLaughlin	15:15	31:53	30:42	29:38	30:07	29:40	29:38
Eddie Billings	16:14	33:15	32:43	32:21	31:59	31:30	31:30
Michael Kipkosgei	14:18	29:45	30:03	29:58	29:03	29:12	29:03
Martin Stoughlin	16:02	32:12	31:47	31:53	31:17	30:12	30:12
Brent Doerhoffer	16:21	33:15	33:08	31:42	31:42	31:17	31:17
Asefe Bekele	14:47	30:08	29:31	29:38	29:12	28:58	28:58
Tom Brinkston	15:37	32:29	31:12	DNR	31:20	DNR	31:12
Mark Berkshire	16:17	DNR	32:28	32:17	32:23	31:38	31:38
Frank Lemit	15:40	32:10	32:28	31:55	31:17	31:23	31:17
Robert Ngugi	15:03	32:00	31:37	31:23	DNR	30:10	30:10
Mekonnen Demissie	14:32	30:04	29:36	29:03	28:51	28:58	28:51
Dereje Tesfaye	15:21	31:54	31:54	31:37	31:43	31:10	31:10
Mel Stein	15:28	31:58	31:47	31:53	31:22	30:36	30:36
Moses Kigen	14:58	30:45	30:30	30:27	29:57	29:32	29:32
Keegan O'Malley	14:54	30:03	30:00	29:43	29:12	DNR	29:12
Tim Williams	15:08	31:55	32:10	31:08	31:11	31:09	31:08

The acronym PR represents the fastest time that the athlete ran and it stands for personal record. DNR=Did not run (typically due to injury)

Figure 6.1. Sample problem statement. From *Statistics for Kids* (pp. 25–26), by S. Chamberlin, 2013, New York, NY: Taylor & Francis. Copyright 2013 by Taylor & Francis. Reprinted with permission.

dren know much more than given credit for and are capable of being challenged to greater levels and areas than previous curricula acknowledge.

Project M^3, similar to its peer Project M^2, was designed with gifted and talented students in mind foremost (Gavin, Casa, Adelson, Carroll, & Sheffield, 2009). Moreover, Project M^3, the first of the two curricula designed, was created using best practices from mathematics education and gifted education. Project M^3 is comprised of 12 units (four each for grades 3, 4, and 5). Unlike Project M^2, which has a focus on geometry and measurement, Project M^3 is interdisciplinary in that it incorporates important ideas from all content areas of mathematics (i.e., algebra, data analysis and probability, geometry, measurement, and number and operations; NCTM, 2000). Also of importance in Project M^3 are teacher suggestions for addressing the classroom climate and the fact that the curriculum was pilot tested with underserved gifted students (Gavin et al., 2009). Moreover, the National Association for Gifted Children recognized several units from Project M^3 as exemplary.

Teachers and stakeholders in gifted education are strongly encouraged to seek Projects M^2 and M^3 for stand-alone curricula, as they are comprised of challenging activities designed specifically to develop advanced learners in mathematics. In fact, they may hold the greatest promise for narrowing the gap between procedural learning and conceptual learning. This gap is one that appears to separate the performance of many Western countries from peers in Southeast Asia in Trends in International Mathematics and Science Study (TIMMS; 2008) data.

The Ideal Curriculum

Readers may be left wondering, "What curriculum will best serve the gifted students with which I work?" Frankly, there is no single answer to this question because each approach fulfills a varied niche from the other approaches and not all classrooms are identical. However, a combination of the four curricula discussed will likely provide the most comprehensive approach for developing gifted mathematicians to function ahead of peers in STEM disciplines. As an example, very young gifted learners could start with the Building Blocks© curriculum that differentiates to their early developmental needs. Young mathematicians can work with this curriculum, and it will help them develop conceptual learning at an early age (pre-K–1). As they progress through kindergarten, such learners can move into Project M^2 and use Building Blocks© as a supplementary curriculum. Near grade 3, learners could transform to the Project M^3 curriculum to realize a more comprehensive overview of mathematics. In the upper years of elementary grades (e.g., grades 4 and 5), young learners can begin to invest serious time solving MEAs. In total, this approach appears to be the single best and most comprehensive approach that can be offered to aspiring mathematicians because all four curricula offer best practices in mathematics education and gifted education, and Project M^2 and Building Blocks© have also taken into consideration best practices by the National Association for the Education of Young Children (NAEYC). Moreover, the comprehensive nature of the four curricula hold great promise for developing mathematicians for multidisciplinary expectations, such as those that will be expected in STEM disciplines.

Discussion Questions

1. Given what you learned in this chapter, what is the best possible learning approach for your mathematics students?
2. Relative to the curriculum characteristics outlined in this chapter, what characteristics in the curriculum you are using engender learning in STEM?
3. How can you change your instructional approach to facilitate conceptual learning in mathematics?
4. Is acceleration or enrichment the preferred approach to facilitating learning among GT mathematics students?

Suggested Resources

▷ Building Blocks©: http://gse.buffalo.edu/org/buildingblocks/
▷ Project M^2: http://projectm2.uconn.edu/
▷ Project M^3: http://www.gifted.uconn.edu/projectm3/
▷ Model-Eliciting Activities: http://c.ymcdn.com/sites/www.amatyc. org/resource/resmgr/2009_conference_proceedings/delmas1.pdf

References

Boaler, J. (2002). Mathematical modeling and new theories of learning. *Teaching Mathematics and its Applications, 3,* 121–128.

Chamberlin, S. A. (2002). Analysis of interest during and after model-eliciting activities: A comparison of gifted and general population students (Doctoral dissertation, Purdue University, 2002). *Dissertation Abstracts International, 64,* 23–79.

Chamberlin, S. A. (2008). What is problem solving in the mathematics classroom? *Philosophy of Mathematics Education, 23,* 1–25.

Chamberlin, S. A. (2013). *Statistics for kids: Model-eliciting activities to investigate concepts in statistics.* Waco, TX: Prufrock Press.

Chamberlin, S. A. (in press). *Using model-eliciting activities to investigate concepts in probability.* Waco, TX: Prufrock Press.

Clements, D. H., & Sarama, J. (2013). *Building Blocks, Volumes 1 and 2.* Columbus, OH: McGraw-Hill Education.

Colangelo, N., Assouline, S., & Gross, M. U. M. (2004). *A nation deceived: How schools hold back America's brightest students.* Iowa City: University of Iowa, The Connie Belin and Jacqueline N. Blank International Center for Gifted Education and Talent Development.

Coxbill, E., Chamberlin, S. A., & Weatherford, J. (2013). Using Model-Eliciting Activities as a tool to identify creatively gifted elementary mathematics students. *Journal for the Education of the Gifted, 37,* 176–197.

Facione, P. A. (1990). *Critical thinking: A statement of expert consensus for purposes of educational assessment and instruction. Research findings and recommendations.* Newark, D: American Philosophical Association. (ERIC Document Reproduction Services No. ED315423).

Gavin, M. K., Casa, T. M., Adelson, J. L., Carroll, S. R., & Sheffield, L. J. (2009). The impact of advanced curriculum on the achievement of mathematically promising elementary students. *Gifted Child Quarterly, 53,* 188–202.

Gavin, M. K., Casa, T. M., Adelson, J. L., & Firmender, J. M. (2013). The impact of advanced geometry and measurement units on the achievement of grade 2 students. *Journal for Research in Mathematics Education, 44,* 478–510.

Gavin, M. K., Casa, T. M., Firmender, J. M., & Carroll, S. R. (2013). The impact of advanced geometry and measurement units on the mathematics achievement of first-grade students. *Gifted Child Quarterly, 57,* 71–84.

Good, T. E., & Brophy, J. E. (1986). *Educational Psychology: A Realistic Approach* (3rd ed.). New York, NY: Longman Publishing.

Hiebert, J., Carpenter, T. P., Fennema, E., Fuson, K. C., Wearne, D., Murray, . . . Human, P. (2000). *Making sense: Teaching and learning mathematics with understanding.* Portsmouth, NH: Heinemann.

Kettler, T. (2014). Critical thinking skills among elementary school students: Comparing identified gifted and general education student performance. *Gifted Child Quarterly, 58,* 127–136.

Köhler, H. (1997). Acting artist-like in the classroom, *International Reviews on Mathematical Education, 29,* 88–93.

Krutetskii, V. A. (1976). *The Psychology of Mathematical Abilities in Schoolchildren* (translated by J. Teller). Chicago, IL: The University of Chicago Press.

Lesh, R., Hoover, M., Hole, B., Kelly, A., & Post, T. (2000). Principles for developing thought-revealing activities for students and teachers. In A. Kelly & R. Lesh (Eds.), *The handbook of research design in mathematics and science education* (pp. 591–646). Hillsdale, NJ: Lawrence Erlbaum and Associates.

Lesh, R. A., & Johnson, H. (1976). Models and applications as advanced organizers. *Journal for Research in Mathematics Education, 7,* 5–81.

Mayer, R. E., & Hegarty, M. (1996). *The process of understanding mathematical problems.* In R. J. Sternberg & T. Ben-Zeev (Eds.), *The nature of mathematical thinking* (pp. 29–54). Mahwah, NJ: Lawrence Erlbaum Associates.

National Council of Supervisors of Mathematics. (1977). Position paper on basic skills. *Arithmetic Teacher, 25,* 19–22.

National Council of Teachers of Mathematics. (1980). *An agenda for action: Recommendations for school mathematics for the 1980s.* Reston, VA: Author.

National Council of Teachers of Mathematics. (1989). *Curriculum and evaluation standards for School Mathematics.* Reston, VA: Author.

National Council of Teachers of Mathematics. (2000). *Principles and standards for school mathematics.* Reston, VA: Author.

National Governors Association Center for Best Practices, & Council of Chief State School Officers. (2010). Common Core State Standards for Mathematics. Washington, DC: Authors.

Richland, L. E., Stigler, J. W., & Holyoak, K. J. (2012). Teaching the conceptual structure of mathematics. *Educational Psychologist, 47,* 189–203.

Rickansrud, K. M. (2011). The impact of the math enrichment program on student achievement (Doctoral dissertation). (ERIC Document Reproduction Services No. ED525318).

Robelen, E. W. (2010). Obama plays cheerleader for STEM. *Education Week, 30,* 1–2.

Trends in International Mathematics and Science Study. (2008). *TIMMS 2007 Results.* Retrieved from https://nces.ed.gov/pubs2009/2009001.pdf

CHAPTER 7

M is for Mathematics at the Secondary Level

Eric L. Mann, Ph.D.,
& Scott A. Chamberlin, Ph.D.

For gifted mathematics students, the transition from elementary school to secondary (middle and high school) often brings a change in the opportunities and challenges offered. If students were fortunate to enjoy the support of a gifted and talented program in elementary school that met their needs, they may find that their differentiated mathematics options at the secondary level are limited where similarly dedicated programs are less commonplace. Acceleration, Advanced Placement courses, or the International Baccalaureate program are the most typical instructional approaches at the secondary level, and each approach has merit but, just as one approach seldom meets the needs of all students in the younger grades, no one approach is sufficient in the secondary grades. After a brief overview of the learning expectations in the Common Core State Standards for Mathematics (CCSS-M), developed under the leadership of the National Governors Association and the Council of Chief State School Officers (2010a), with respect to the needs of gifted learners, the strengths and weakness of each instructional approach most commonly used at the secondary level will be explored, as will recommendations to differentiate and enrich all curriculum programs. The chapter concludes with a discussion of the role math-

 DOI: 10.4324/9781003238218-8

ematics plays in the development of STEM (science, technology, engineering, and mathematics) talent and creativity at the secondary level.

Learning Expectations and Curriculum Standards

The effort to develop a set of model academic standards grounded in rigorous content and the application (as opposed to the dissemination) of knowledge though the use of higher-order thinking skills predates the CCSS-M. In 1894, the National Education Association (NEA) at Harvard University convened the Committee of Ten. In their report, the committee recommended the following with respect to secondary mathematics education:

> The method of teaching should be throughout objective, and such as to call into exercise the pupil's mental activity. The text-books should be subordinate to the living teacher. The illustrations and problems should, so far as possible, be drawn from familiar objects, and the scholar himself should be encouraged to devise as many as he can. So far as possible, rules should be derived inductively, instead of being stated dogmatically. On this system the rules will come at the end, rather than at the beginning, of a subject. (p. 105)

Many similarities can be found between this document and the standards for mathematical practice in the CCSS-M that stress, in part, the need for students to make sense of problems, construct viable arguments, use tools appropriately and strategically and look for structure and regularity in mathematics rather than just learning rules and procedure. Johnsen and Sheffield (2012) recommended the addition of another standard of mathematical practice to the eight in the CCSS-M: "solve problems in novel ways and pose new mathematical questions of interest to investigate" (p. 16). Paralleling these efforts, the Partnership for 21st Century Skills, in collaboration the Mathematical Association of America and the National Council of Teachers of Mathematics developed a 21st Century Skills Map that recognizes the learning and innovation skills needed to prepare our students for the complex life and work environments in which they will find themselves (Partnership for 21st Century Skills, 2011).

The message here is that it is no longer (and never really has been) sufficient to simply offer students content without context or opportunities to move beyond the mastery of the textbook. Rather students need and deserve "challenging classroom environments and curricula that develop and nurture mathematical talent, creativity, and zeal" (National Council of Supervisors of Mathematics, 2012, p. 1). A step in this direction is Appendix A of the CCSS-M (National Governors Association & Council of Chief State School Officers, 2010b) that addresses the design of high school mathematics courses offering pathways (course recommendations) for both the traditional approach seen in the United States and an integrated approach more typically found internationally. "Compacted" versions of these pathways are also offered. Unlike curriculum compacting (Reis & Renzulli, 2005), where previously mastered content is eliminated to gain opportunities to enrich or accelerate the student's study of mathematics, the CCSS-M include all the content in the "noncompacted" versions offered at an accelerated pace. Worth noting here is that all four pathways presume a 3-year high school mathematics course of study preparing students for entry into courses in higher level mathematics, such as precalculus, calculus, or advanced statistics (National Governors Association & Council of Chief State School Officers, 2010b). A mathematics program for gifted and talented students that only focuses the content, practices, and dispositions in the CCSS-M is insufficient; they are capable of so much more.

Curricular Options

Acceleration

Acceleration is one of the cornerstones of exemplary programs for gifted students but often is done by offering early access to higher level courses without tailoring the course for the student (National Association for Gifted Children [NAGC], 2004). The opening pages of *A Nation Deceived: How Schools Hold Back America's Brightest Students* (Colangelo, Assouline, & Gross, 2004) defined acceleration as ". . . an intervention that moves students through an educational program at rates faster, or at younger ages, than typical. It means **matching the level, complexity, and pace of the curriculum to the readiness and motivation of the student**" (p. xi; emphasis added). Junter and Sriraman (2011) stressed the importance of including curriculum compacting (trimming out repetitive tasks) and differentiation in addition to acceleration—it's more than just moving a student several grades level higher on the curriculum ladder. Depending on the implementation of the CCSS-M, compacted pathway students may find these courses simply move at a faster pace. For example, the description of the course of study for the CCSS-M accelerated traditional pathway in grade 7 opens with, "This course differs from the non-accelerated 7th grade course in that it contains content from 8th grade. While coherence is retained . . . the additional content . . . demands a faster pace for instruction and learning" (National Governors Association & Council of Chief State School Officers, 2010b, p. 92). Therefore, this course offers little for the seventh-grade Talent Search student who may have completed Algebra I in sixth grade (typically ninth-grade content,) as it is a step backward in the progression of courses and content outlined in the CCSS-M pathways.

Although the decision to accelerate should be informed by multiple data sources for each individual student, often such decisions are driven by other factors. For example, a recent joint report from the National Council of Teachers of Mathematics and the Mathematical Association of America (as cited in Bressoud, Camp, & Teague, 2012) on students completing precalculus or calculus in high school noted that:

> What the members of the mathematical community—especially those in the Mathematical Association of America (MAA) and the National Council of Teachers of Mathematics (NCTM)—have known for a long time is that the pump that is pushing more students into more advanced mathematics ever earlier is not just ineffective: It is counter-productive. Too many students are moving too fast through preliminary courses so that they can get calculus onto their high school transcripts. The result is that even if they are able to pass high school calculus, they have established an inadequate foundation on which to build the mathematical knowledge required for a STEM career. (p.2)

There is compelling evidence that, when delivered thoughtfully and purposefully, acceleration is an effective strategy to meet the needs of some children (see Colangelo, Assouline, & Gross, 2004), but it is not a sufficient or global strategy for a secondary gifted program. "Acceleration comes down to one child, one family, one situation" (Colangelo, Assouline, & Gross, 2004, p. xi).

Advanced Placement Courses

The College Board's Advanced Placement (AP) program continues to see significant growth. In 2008, Hertberg-Davis and Callahan summarized their findings of students' perceptions of AP courses and the International Baccalaureate (IB) program, reporting more than one million students were involved in AP programs in 2005. Data for 2013 showed an increase in AP enrollment of 205% (1,081,102 in 2005 versus 2,218,578 in 2013; College Board, 2013). Although consistently receiving favorable reviews in gifted education literature (see Hertberg-Davis, Callahan, & Kyburg, 2006), AP courses in many secondary schools constitute the only gifted education program option (Hertberg-Davis & Callahan, 2008; Hertberg-Davis, Callahan, & Rykurg). Emphasis on improving access to AP courses to strengthen K–12 STEM education comes from both the public and private sectors. A leader in this effort is the National Math and Science Initiative (see http://www.nms.org), which seeks to increase the number of students taking and passing AP math, science, and English examinations and expand access to underrepresented populations. However, although increasing access to AP courses draws significant resources, offering AP courses as *the solution* to meet the educational needs of all gifted and talented students is at best a limited approach.

AP courses strive to engage students in coursework equivalent to introductory courses at the college level and are designed by committees involving experienced AP teachers and college faculty (National Research Council, 2002). Yet Callahan (2003) found that:

> No systematic research documents the equivalence of AP and IB Programs to college courses, and the literature does not present evidence that exam scores predict success in upper-level college courses or that these courses provide the depth of understanding equivalent to that of introductory college courses (NAS, 2002). Further, in reviewing the mathematics and science curriculum offered in AP and IB Programs, the National Academy panel (NAS, 2002) concluded that there were shortcomings in these curricula in terms of the development of key ideas of the disciplines and metacognitive skills, and delineation of prior knowledge required. (p. xi)

Qualitative studies of gifted students' perceptions of AP and IB programs (see Hertberg-Davis & Callahan, 2008; Hertberg-Davis, Callahan, & Kyburg, 2006) have found that most prefer AP courses in lieu of the general education alternatives: "... an opportunity to escape from the drudgery of less challenging course" (Hertberg-Davis & Callahan, 2008, p. 202). Students "... successful in AP and IB classes represented a largely self-selected group, fitting a student profile of long-term success, self-motivation, drive to succeed and conformity to school expectations. While this profile accurately describes many gifted secondary students, it also excludes many ..." (p. 70). Although AP was perceived as more challenging and a higher quality learning experience, some students found the rigidity and "one-size-fits-all" approach was a poor fit for the way they best learned: "... students who did not fit the 'AP/IB mold' of long-time school success did not perceive the one-size-fits-all, fast-paced courses to be a good fit for their needs" (Hertberg-Davis & Callahan, 2008, p. v).

The National Research Council (2002) voiced concerns over the AP curriculum's balance between breadth and depth of coverage, noting that with the notable exception of the AP calculus syllabus, AP curriculum is "... incompatible with a curriculum designed to foster deep conceptual understanding" (p. 176). The breadth of material to be covered often results in instruction focused on "... presenting information and teaching problem types, rather than facilitating active, student involvement in learning" (p. 179), an instructional approach inconsistent with NAGC's (2010) Pre-K–Grade 12 Gifted Education Programming Standards.

The educational philosophy in the AP Calculus Course Teacher's guide is "... primarily concerned with developing the students' understanding of the concepts of calculus and providing experience with its methods and applications ... the connections among these representations also are important .... broad concepts and widely applicable methods are emphasized" (Howell, 2007, p.4). The Advanced Academics Curriculum Evaluation System (AACES) Task Force formed by the Texas Education Agency (TEA; 2007) reviewed AP programs for use with gifted students, rating courses in four areas: content, process, product, and affect. The AP Calculus program's highest scores were in content and process but lacked especially in the affect domain (see Tables 7.1 and 7.2). The TEA (2007) Task Force found that:

> Some items in the Product domain indicated potential for differentiation, but the Affect domain yielded no overt strategies for meeting the needs of gifted learners. Furthermore, career connections and varied resources were not explicit in the course materials. Including connections to math-related career fields and providing more varied resources and manipulatives could increase gifted students' understanding in the content and process domains as well as develop personal interests

TABLE 7.1
Average Scores from the ACCE Task Force Evaluation

Category	Calculus AB	AP Biology	English Language & Composition	U.S. History
Content	1.9	2.6	2.6	2.1
Process	1.9	2.7	2.7	2.6
Product	1.5	2.3	2.7	1.7
Affect	1.0	1.7	2.7	2.7

1: Indicates that the element was not present
2: Suggests that there is potential in the lesson, but the inclusion of the element is not overt
3: Signifies that the element is found

Data from *G/T teacher toolkit II: A set of resources for teachers of G/T, AP, and Pre-AP classes* (Retrieved from http://www.texaspsp.org/toolkit2/Toolkit2.html), by Texas Education Agency, 2007, Austin, TX: Author. Copyright 2007 by Texas Education Agency. Adapted with permission.

TABLE 7.2
Selected Items from AACE Task Force Findings

Category	Item	Descriptor
Content	4	Allows for in-depth learning of a self-selected topic
	9	Offers opportunities for students to engage in activities aligned with students' individual strengths, preferences or interests
	11	Allows for the acceleration of content in an area of strength
Process	1	Develops independent or self-directed study skills
	2	Focuses on open-ended tasks
Product	1	Develops products that challenge existing ideas and produce "new" ideas
	2	Develops products that use new techniques, materials, and forms
Affect	1	Encourages the development of self-understanding
	2	Encourages growth and change in the student's abilities and personal outlooks
	3	Includes mentors/tutors who share common interests and talents with students

Note. AP Calculus AB Items Scored at 1.0, Not Present

Data from *G/T teacher toolkit II: A set of resources for teachers of G/T, AP, and Pre-AP classes* (Retrieved from http://www.texaspsp.org/toolkit2/Toolkit2.html), by Texas Education Agency, 2007, Austin, TX: Author. Copyright 2007 by Texas Education Agency. Adapted with permission.

as noted in the affective domain. Further recommendations by the Task Force mathematics specialists included providing gifted learners with alternative upper-level mathematics courses which have a greater potential for differentiation strategies, such as AP Statistics. (p. 14)

AP courses help fill a void in the educational programming offered for those gifted and talented students not only seeking challenge but also the opportunity to be with students who share like interests and with a teacher whom they view as knowledgeable and dedicated. As an opt-in course rather than a required one, students are there because they see value in participating. Hertberg-Davis and Callahan's (2008) study reported an overwhelming preference for the learning environment in AP courses among the gifted and talented students participating in their study attributed to two key factors: "(a) the opportunity to learn with students of similar abilities, motivation and academic interests and (b) the adultlike relationships they had with their AP and IB teachers" (p. 204). At the same time, the nature of the program excludes students who have the intellectual capacity and desire to succeed but may not have had the requisite experiences. Often, the students are from underrepresented populations resulting in "relativity homogeneous classrooms" (Hertberg-Davis & Callahan, p. 206). In our efforts to expand the STEM talent pool, motions are needed to remove the barriers (real or perceived) that cause student to opt-out rather than opt-in.

AP courses offer an alternative to the general education classroom that is appropriate for some students. Comparable to acceleration in their approach to offer advanced curriculum with a significant infrastructure in place to provide curriculum materials, prepare teachers, assess students, and reward accomplishments, AP courses have a place in a continuum of services offered the secondary gifted and talented student. Recommendations from the Hertberg-Davis and Callahan (2008) article are particularly relevant here and track closely with the National Research Council's (2002) concern over program deficiencies and United States Department of Education's (USDOE) efforts to " . . . increase the number of gifted and talented students from underrepresented groups who, through gifted and talented education programs, perform at high levels of academic achievement" (USDOE, 2014, p.5). In particular, it is necessary to develop in AP teachers the skills needed to differentiate curriculum and vary instructional strategies to meet unique needs—something that has been recognized in some states. In Georgia, for example, in order for an AP course to count toward meeting gifted programming requirements, not only must the teacher have the appropriate content area certification and College Board AP training for the specific course taught, but also " . . . must have completed a 10 clock hour professional development course in characteristics of gifted learners and curriculum differentiation for gifted students" or hold a current state-issued gifted endorsement (Georgia Resource Manual for Gifted Education Services, 2014–2015, p. 14). These requirements also apply to International Baccalaureate (IB) Diploma Courses used to meet the needs of gifted students. Readers seeking more information on differentiated instruction for gifted secondary math students, should see *Using the Common Core State Standards for Mathematics With Gifted and Advanced Learners* (Johnsen & Sheffield, 2013) and *More Good Questions: Great Ways to Differentiate Secondary Mathematics Instruction* (Small & Lin, 2010).

At present, resources are sparse to support differentiation in AP classrooms in part because of the breadth of content to be covered in the weeks leading up to the AP exam. In his discussion of *The Calculus Trap,* Rusczyk (2015) wrote that ". . . the gifted, interested student should be exposed to mathematics outside the core curriculum, because the standard curriculum is not designed for the top students. This is even, if not especially, true for the core calculus curriculum found at most high schools, community colleges, and universities" (p.1). Rather than focusing on coverage and adding new mathematical tools to a gifted student's skillset, Rusczyk (2015) advocates for an approach that allows our students to use the tools they have and learn how to apply them to increasing complex problems. A few suggestions for ways this might be accomplished are offered in a later section in this chapter on curriculum enrichment.

International Baccalaureate Courses

Developed in the1960s, the IB Diploma Program is "a comprehensive integrated precollege curricula for highly motivated student in the last 2 years of high school" (National Research Council, 2002, p. 83). As a program designed to prepare students for college through an integrated, multidisciplinary course of study rather than to deliver college-level courses to high school students, the IB program may offer teachers and students more flexibility than available from a menu of AP course options. The National Research Council (2002) noted that while the IB syllabi, requirements, and assessments lead to less variability in content than corresponding AP courses, they also offer alternative teaching strategies in the accompanying teaching notes that include the use of informal investigations, problem solving without the use of explicit formula, and suggestions to help see connections between concepts. Unlike AP's offering of discrete courses, the IB program is compressive program of study spanning six subject areas with courses offered at the Higher Level and Standard Level. In addition to their subject level coursework, IB Diploma candidates complete three additional requirements. A 2-year course, Theory of Knowledge, seeks to engage students in critical reflection and "... centers on a series of questions, including those designed to help students understand the nature of knowledge in mathematics and sciences" (National Research Council, 2002, p. 85). Students also complete an original research project. The Extended Essay requirement allows for in-depth study of a topic the student chooses and emphasizes the development of communication skills needed for college. The third requirement, Creative, Action and Service, encourages student involvement outside of the classroom.

Four mathematics courses are offered in the IB program at varying levels. A description of the stated goals from them is provided below:

▷   develop mathematical knowledge, concepts, and principles;

▷   develop logical, critical, and creative thinking; and

▷   employ and refine [students'] powers of abstraction and generalization.

Students are also encouraged to appreciate the international dimensions of mathematics and the multiplicity of its cultural and historical perspectives. (International Baccalaureate Organization, 2005–2014).

Course offerings in the IB program differ from AP Calculus in that they cover a range of advanced mathematical concepts and topics, more along the lines of integrated pathway discussed in Appendix A of the CCSS-M (National Governors Association & Council of Chief State School Officers, 2010b). For example:

The IB Mathematics HL course is taught over two years and includes as core material a substantial amount of calculus, roughly equivalent to the AP Calculus AB course, as well as substantial treatment of probability, algebra and trigonometry, complex numbers, mathematical induction, vectors, and matrices. In addition, the HL curriculum has optional units, one of which is taught in addition to the common core. There are units on abstract algebra, graphs and trees, statistics, analysis and approximation, and Euclidean geometry and conic sections. With the analysis and approximation option, the calculus coverage becomes roughly equivalent to that of AP Calculus BC. (National Research Council, 2002, p. 500).

For students seeking challenge and a community of peers, the IB program offers an immersive opportunity. Many similarities can be found between IB program and various gifted education models (e.g., varying levels of challenge, opportunities to pursue areas of interest, emphasis on critical and creative thinking, etc.) In 2008, the Texas Association for Gifted and Talented concluded the IB program " . . . offers avenues of inquiry, appropriate assessment measures, and opportunities to meet the affective needs and strengths for gifted students . . . supply sound pedagogical and instructional components that facilitate learning for the gifted" (Boswell, 2008, p. 2). Yet many of the same concerns raised in the earlier discussion of AP programming are relevant here as well. Hertberg-Davis and Callahan's (2008) study grew out of the many unanswered questions on the appropriateness of either program for gifted secondary students. In an answer to a parent's question on the choice between AP and IB on Duke University's Talent Identification Program website, Callahan's (2006) final paragraph is relevant here.

> Four mathematics courses are offered in the IB program at varying levels. A description of the stated goals from them is provided below:
> ▷ develop mathematical knowledge, concepts, and principles;
> ▷ develop logical, critical, and creative thinking; and
> ▷ employ and refine [students'] powers of abstraction and generalization.

> Either AP or IB is likely to provide more advanced content and greater challenge than other courses, as well as the opportunity to earn college credit, and success in these programs is highly regarded by college admissions offices. However, the true value of these learning opportunities depends on the fit between your daughter's learning style, motivation, and preparation for the challenges offered by AP and IB. (para. 9)

So as with acceleration and AP courses, no one approach will meet the needs of all children.

Suggestions for Curriculum Enrichment

History of the Discipline

An understanding of the history of mathematics adds the human touch to the concepts students encounter in the classroom. Students struggling to understand the concepts of negative numbers may enjoy understanding that concept was a controversial one compelling one French scholar to attribute the "failure of the teaching of mathematics in France to the admission of negative quantities" and declaring that such mental aberrations could prevent gifted minds from studying mathematics (Busset, 1843/2010, p. 47). Berlinghoff and Gouvêa's (2004) *Math Through the Ages: A Gentle History for Teachers and Others* and Posamentier's (2003) *Math Wonders to Inspire Teachers and Students* are excellent sources of topics to add richness and depth to the study of mathematics. For the younger student, Pappas's *Mathematical Scandals* (1997/2002) and *The Joy of Mathematics: Discovering the World Around You* (1989/2001) offer intriguing and thought-provoking insights into mathematical history.

Developing Mathematical Thinking

Persistence in problem solving is an essential skill. In his March 2010 TEDTalk, *Math Class Needs a Makeover,* Dan Meyer coined the phrase "impatient problem solvers," and although he attributed part of the problem to the world in which we find ourselves, he asserted that the curriculum approach in many mathematics classrooms contributes to the problem. Thirty problems of homework a night or timed tests may develop endurance but does little to develop the persistence needed for students to tackle difficult problems. In her book *Faster Isn't Smarter: Messages About Math, Teaching, and Learning in the 21st Century,* NCTM Past President Cathy Seeley (2009) wrote of the potential damaging effects timed tests and competitive mathematics may have on many students' confidence and willingness to tackle new problems. Fiori (2007) wrote, "Most of my mathematician friends and I are only able to solve about two problems a year—if we're lucky! Tell a mathematician you've solved even five problems in a single day, and the first thing she will think is, 'They must not have been very interesting problems'" (p. 695). Finding ways to engage students in deep mathematical thinking beyond what they experience in math competitions, where speed and rapid insight are rewarded, is necessary, but so, too, are ways to reward students whose talents lie more in contemplative realms. There are many approaches to developing mathematical thinkers; however, an exhaustive review is unfortunately beyond the scope of this chapter. What follows is an introduction to three such instructional programs: Britain's Masterclasses, Russian Mathematical Circles, and Bay Area Mathematical Adventures.

Mathematics Masterclasses were introduced by George Porter (1997), the first president of the National Association for Gifted Children in the United Kingdom, to help meet the needs of talented mathematics students. Sewell (1997) offered an insight into the nature of these classes in his text *Mathematics Masterclasses: Stretching the Imagination*. In the foreword Porter described students involved in these classes as passing their A-level exams (completing their secondary education) in their early teens and graduating from university before age 20, yet the text is suitable for 13 year olds with a bit of mathematical curiosity. At the time of the book's printing, Masterclasses were reaching over 250,000 students a year in Britain. The text contains 12 chapters as samples of classes offered, some with a more traditional focus, such as a chapter on Pythagoeran number triples, and some with a more integrated STEM focus, such a chapter on Plato, polyhedra, and weather forecasting.

From Russia, we benefit from an almost century-long tradition of mathematical circles, described in Formin, Genkin, and Itenberg's (1996) *Mathematical Circles: Russian Experience*. Theses circles involved students, teachers, and mathematicians working together to solve problems based on the premise that " . . . studying mathematics can generate the same enthusiasm as playing a team sport, without necessarily being competitive" (p. vii). Some of the chapters supplement topics addressed in standards-based courses, and others enrich the mathematical experience by offering topics not often encountered until upper-level undergraduate study. Problems are sequenced to engage students in progressively more challenging tasks building the experience to develop problem-solving skills.

The Bay Area Mathematical Adventures (BAMA) talks seek to expose high school age students to the beauties of mathematics through published lectures. *Mathematical Adventures for Students and Amateurs* (Hayes & Shubin, 2004) is a collection of these talks. A second book, *Expeditions in Mathematics,* (Shubin, Hayes, & Alexanderson, 2011) adds additional BAMA lectures. Reading about lectures may not fit the view of mathematics as a process of active problem solving; however, within this context students get the opportunity to a glimpse into aspects of mathematics beyond the topics they encounter in a typical high school classroom—an experience that changes mathematics from a series of rules and procedures to master into a creative, exploratory act. A nice complement to these books is the YouTube video series Curriculum Inspirations, in which there are more than 100 videos created as a multimedia experience for secondary students (see http://www.youtube.com/playlist?list=PLevtNOOa6SZXV-JvtROAFCC0oYt0ySTSo4). Hosted by Dr. James Tanton, each video is based on a problem from American Mathematics Competitions. For example, in the video *Angles in a Star*, Dr. Tanton first presents a geometry problem than can be solved by employing properties of angles and intersecting lines and reframes it into an exploration of angles as a measure of turning and of the sum of the angles in the point of the stars. Extending the problem in this way fits well with the recommendation from Johnsen and Sheffield (2012), discussed earlier in this chapter.

> In order to support mathematically advanced students and to develop students who have the expertise, perseverance, creativity, and willingness to take risks and recover from failure, which is necessary for them to become mathematics innovators, we propose that a ninth Standard for Mathematical Practice be added for the development of promising mathematics students—a standard on mathematical creativity and innovation: *Solve problems in novel ways and pose new mathematical questions of interest to investigate.* (pp. 15–6)

Masterclasses, Math Circles, and BAMA are shared here, not as an exhaustive list of enrichment materials but rather as a small sampling of the richness of the resources of materials that are available for classroom use. As students move on to college and into the world of work, they will encounter real-world problems that are ill formed and that require a variety of methods and skills to interpret and solve. By sharing with them the beauty and creative nature of mathematics through the works and words of mathematicians and providing opportunities to design and answer their own problems, educators will better prepare learners for the world they will encounter.

Mathematics Education and STEM

Almarode, Subotnik, Crowe, Tai, Lee, and Nowlin (2014) investigated the role of self-efficacy and maintenance of interest in the STEM disciplines from students participating in specialized high schools and talent search programs. Their study suggested that maintaining, and better yet, intensifying an individual's interest in STEM is " . . . strongly and positively associated with their persistence and earning an undergraduate degree in STEM" (p. 327). Although this is certainly possible in a general education classroom, the student perceptions noted in Hertberg-Davis and Callahan's (2008) study suggest this rarely happens.

Mathematical talents are best developed and nurtured in an environment that embraces passion and creativity (National Council of Supervisors of Mathematics, 2012). Yet we know that:

> . . . performance in mathematics courses, often does not effectively predict who will succeed as a mathematician. The prediction failure occurs due to the fact that in math, as in most other fields, one can get away with good analytical but weak creative thinking until one reaches the highest levels of mathematics. (Sternberg, 1996, p. 313)

If an individual with creative potential in mathematics or a related STEM discipline is not provided an outlet for that creativity until graduate or postgraduate work, maintaining an interest in mathematics is difficult at best. Offering advanced curricular options to gifted and talented STEM secondary students without grounding those experiences in the mathematical practices, or without experiencing the challenges and rewards of doing mathematics as a practicing professional—that is, as a doer rather than as a consumer of mathematics—then talent development opportunities are lost.

Mathematics—this may surprise you or shock you some—is never deductive in its creation. The mathematician at work makes vague guesses, visualizes broad generalizations, and jumps to unwarranted conclusions. He arranges and rearranges his ideas, and he becomes convinced of their truth long before he can write down a logical proof. The conviction is not likely to come early—it usually comes after many attempts, many failures, many discouragements, many false starts. (Halmos, 1968, p. 286)

> " . . . performance in mathematics courses, often does not effectively predict who will succeed as a mathematician. The prediction failure occurs due to the fact that in math, as in most other fields, one can get away with good analytical but weak creative thinking until one reaches the highest levels of mathematics" (Sternberg, 1996, p. 313).

There are few surprises in solving problems with known solutions, often with answers provided in the back of the book or the teacher's edition. Those with gifts and talents in mathematics deserve the opportunity to make false starts, regroup, and try again, and along way be delighted with the surprises they discover. Those responsible for providing for the needs of these students deserve the same.

Discussion Questions

1. If you had the opportunity to create a course for your gifted mathematics students, what mix of programming options outlined in this chapter might best serve them?

2. Developing persistence in problem solving is challenging, and not quickly seeing the "right" way to solve a problem is often frustrating, especially for gifted students who grow accustomed to rapid success. How might you create a learning environment that values persistence in today's environment with the emphasis on content and skill acquisition, accountability, and test scores?

3. If acceleration and advanced coursework were replaced as the predominate method of providing services to gifted students in secondary math classrooms, what new courses might be offered to deepen student conceptual understanding of mathematics?

4. Recent emphasis on STEM education suggests a more integrated teaching approach to help students see the connections among the disciplines—an approach that appears to value applied mathematics and career readiness. Does such an approach improve or impoverish the learning experiences of our gifted secondary math students?

References

Almarode, J. T., Subotnik, R. F. Crowe, E., Tai, R. H. Lee, G. M., & Nowlin, F. (2014). Specialized high schools and talent search programs: Incubators for adolescents with high ability in STEM disciplines. *Journal of Advanced Academics, 25,* 307–331.

Berlinghoff, W. P. & Gouvêa, F. Q. (2004). *Math through the ages: A gentle history for teachers and others (expanded edition)*. Washington, DC: Mathematical Association of America.

Boswell, C. (2008). *Gifted learners and International Baccalaureate° primary years, middle years, and diploma programmes.* Austin, TX: Texas Association for Gifted and Talented.

Bressoud, D., Camp, D., & Teague, D. (2012). *Background to the MAA/NCTM statement on calculus.* Reston, VA: NCTM.

Bussett, F. C. (2010). *de L'enseignement des mathematiques dans les colleges.* Whitefish, MT: Kessinger Publishing. [Original work published 1843]

Callahan, C. M. (2003). *Advanced Placement and International Baccalaureate programs for talented students in American high schools: A focus on science and math* (Research Monograph 03176). Storrs: University of Connecticut National Research Center on the Gifted and Talented. Retrieved from http://www.gifted.uconn.edu/nrcgt/reports/rm03176/rm03176.pdf

Callahan, C. M. (2006). Advanced Placement or International Baccalaureate? *Digest of Gifted Research.* Durham, NC: Duke Talent Identification Program. Retrieved from http://tip.duke.edu/node/815

Colangelo, N.. Assouline, S. G., Gross, M. U. M. (2004). *A nation deceived: How schools hold back America's brightest students.* Iowa City: University of Iowa, The Connie Belin & Jacqueline N. Blank International Center for Gifted Education and Talent Development.

College Board. (2013). Annual AP program participation 1956–2013. Retrieved from http://media.collegeboard.com/digitalServices/pdf/research/2013/2013-Annual-Participation.pdf

Fiori, N. (2007). Four practices that math classrooms could do without. *Phi Delta Kappan, 88,* 695–696.

Formin, D., Genkin, S., & Itenberg, I. (1996). *Mathematical circles: Russian experience.* Providence, RI: American Mathematical Society.

Georgia Resource Manual for Gifted Education Services. (2014–2015). Retrieved from http://www.gadoe.org/Curriculum-Instruction-and-Assessment/Curriculum-and-Instruction/Documents/Gifted%20Education/2014-2015-GA-Gifted-Resource-Manual.pdf

Halmos, P. (1968). Mathematics as a creative art. *American Scientist, 56,* 375–389.

Hayes, D. F., & Shubin, T. (Eds.). (2004). *Mathematical adventures for students and amateurs.* Washington, DC: The Mathematical Association of America.

Hertberg-Davis, H., & Callahan, C. M. (2008). A narrow escape: Gifted students' perceptions of advanced placement and international baccalaureate programs. *Gifted Child Quarterly, 52,* 199–216.

Hertberg-Davids, H., Callahan, C. M., & Kyburg, R. M. (2006). *Advanced Placement and International Baccalaureate programs: A "fit" for gifted learners?* (Research Monograph 0622). Storrs: University of Connecticut National Research Center on the Gifted and Talented. Retrieved from http://www.gifted.uconn.edu/nrcgt/hertcall.html

Howell, M. (2007). *AP® calculus teacher's guide.* New York, NY: The College Board. Retrieved from http://apcentral.collegeboard.com/apc/members/repository/ap07_calculus_teachersguide_2.pdf

International Baccalaureate Organization. (2005–2014). *Diploma programme curriculum: Group 5: Mathematics.* Retrieved from http://www.ibo.org/diploma/curriculum/group5/

Johnsen, S. K., & Sheffield, L. J., (2012), *Using the Common Core State Standards for mathematics with gifted and advanced learners.* Waco, TX. Prufrock Press.

Junter, K., & Sriraman, B. (2011). Does high achieving in mathematics = gifted and/or creative in mathematics. In B. Sriraman & K. H. Lee (Eds.), *The elements of creativity and giftedness in mathematics* (pp. 45–65). Rotterdam, The Netherlands: Sense Publisher.

Meyer, D. (2010). Math class needs a makeover. Retrieved from http://www.ted.com/talks/dan_meyer_math_curriculum_makeover?language=en

National Association for Gifted Children. (2004). *Position statement: Acceleration.* Washington, DC: Author. Retrieved from http://www.nagc.org/sites/default/files/Position%20statement/Acceleration%20Position%20statement.pdf

National Association for Gifted Children. (2010). *NAGC Pre-K–Grade 12 gifted education programing standards.* Washington, DC: Author. Retrieved from http://www.nagc.org/sites/default/files/standards/K-12%20programmingstandards.pdf

National Council of Supervisors of Mathematics. (2012). *Improving student achievement in mathematics by expanding opportunities for our most promising students of mathematics.* Denver, CO: Author.

National Educational Association. (1894). *Report of the Committee of Ten on secondary school studies with the reports of the conferences arranged by the committee.* New York, NY: American Book Company.

National Governors Association Center for Best Practices, & Council of Chief State School Officers. (2010a). *Common Core State Standards for mathematics.* Washington, DC: Author.

National Governors Association Center for Best Practices, & Council of Chief State School Officers. (2010b). *Common Core State Standards for mathematics Appendix A: Designing high school mathematics courses based on the Common Core State Standards.* Washington, DC: Author. Retrieved from http://www.corestandards.org/assets/CCSSI_Mathematics_ Appendix_A.pdf

National Research Council. (2002). *Learning and understanding: Improving advanced study of mathematics and science in U.S. high schools.* Washington, DC: National Academy Press.

Pappas, T. (2001). *The joy of mathematics: Discovering mathematics all around you.* San Carlos, CA: Wide World Publishing/Tetra. [Original work published 1989]

Pappas, T. (2002). *Mathematical scandals.* San Carlos, CA: Wide World Publishing/Tetra. [Original work published 1997]

Partnership for 21st Century Skills. (2011). 21st century skills map – math. Washington, DC: Author. Retrieved from http://www.p21.org/storage/ documents/P21_Math_Map.pdf

Porter, G. (1997). *Mathematics masterclasses: Stretching the imagination.* London, England: Oxford University Press.

Posamentier, A. S. (2003). *Math wonders to inspire teachers and students.* Alexandria, VA: ASCD.

Reis, S. M, & Renzulli, J. S. (2005). *Curriculum compacting: An easy start to differentiating for high-potential students.* Waco, TX: Prufrock Press.

Rusczyk. R. (2015). *The calculus trap.* Retrieved from http://www.artof problemsolving.com/Resources/articles.php?page=calculustrap&

Seeley, C. L. (2009). *Faster isn't smarter: Messages about math, teaching, and learning in the 21st Century.* Sausalito, CA: Math Solutions.

Sewell, M. (Ed.). (1997). *Mathematics masterclasses: Stretching the imagination.* London, England: Oxford University Press.

Shubin, T. S., Hayes, D. F., & Alexanderson, G. L. (2011). *Expeditions in mathematics.* Washington, DC: Mathematical Association of America/ Spectrum.

Small, M., & Lin, A. (2010). *More good questions: Great ways to differentiate secondary mathematics instruction.* Reston, VA: National Council of Teachers of Mathematics.

Sternberg, R. J. (1996). What is mathematical thinking? In R. J. Sternberg & T. Ben-Zeev (Eds.), *The nature of mathematical thinking* (pp. 303–318). Mahwah, NJ: Lawrence Erlbaum.

Texas Education Agency. (2007). *G/T teacher toolkit II: A set of resources for teachers of G/T, AP*, and Pre-AP* classes.* Retrieved from http://www. texaspsp.org/toolkit2/documents/gttoolkit2.pdf

U.S. Department of Education. (2014). *Fiscal year 2014 application for new grants under the Javits Gifted and Talented Students Education Program* (CFDA 84.206A). Washington, DC: Author. Retrieved from http://www2.ed.gov/programs/javits/2014-206a.pdf.

CHAPTER 8

Assessing Aptitude and Achievement in STEM Teaching and Learning

Amy L. Sedivy-Benton, Ph.D., Heather A. Olvey, & James P. Van Haneghan, Ph.D.

Introduction

Over the last several years there have been concerns raised by various reports (e.g., President's Council of Advisors on Science and Technology, 2010) that the United States has not been producing enough individuals with expertise in the areas of science, technology, engineering, and mathematics (STEM). Hence, identifying and nurturing the talents of students who are identified as high ability in STEM areas is an important educational and workforce objective. In this chapter, we explore the assessment strategies that can be used to uncover and facilitate STEM learning in students of high ability. We will first explore methods for identifying STEM potential and aptitude. Then, we will discuss assessment practices that can be used to facilitate and measure the development of interest and achievement in STEM areas.

 DOI: 10.4324/9781003238218-9

Identification of Students With Aptitude for STEM

There has been a long history of using standardized assessments for talent identification purposes (e.g., Keating, 1976). Programs out of Johns Hopkins University (http://cty.jhu.edu/talent/schools/students.html) and Duke University (http://tip.duke.edu/) start by examining high student achievement on nationally normed standardized grade-level tests (e.g., the 95th percentile on a test or subtest) and then identify talent further by looking at performance on tests that are above grade level (e.g., the SAT or ACT). The logic is that students will not reach the ceiling on the test and that it will differentiate the talent levels of the students further.

Although this approach is successful in capturing a large number of students who would traditionally succeed in STEM fields, there is evidence that some students are missed in such a process. For example, minority students are not adequately captured through standardized testing (McBee, 2010). Additionally, a number of researchers have presented evidence that students high in spatial ability, an area that is correlated with STEM success, are often missed because the standardized assessments do not tap into this aptitude adequately (Andersen, 2014; Kell, Lubinski, & Benbow, 2013; Mann, 2006; Shea, Lubinski, & Benbow, 2001; Webb, Lubinski, & Benbow, 2007). Many students who are strong in STEM subjects have a high level of spatial ability (McClain & Pfieffer, 2012; Misset, & Brunner, 2013). Although visual-spatial skill can predict success in STEM subject areas (Andersen, 2014; Coxon, 2012; Newcombe, 2010), in practice, educators rarely measure visual-spatial ability or work to develop it in students. Many teachers focus on more common assessments of quantitative and verbal reasoning. Research has found that some students with strengths in spatial reasoning can struggle with verbal challenges and rote memorization, areas that most people find easy to master (Mann, 2006). Further, findings from Project Talent showed that "more than half of the participants in the top 1% of spatial ability were not in the top 3% of math and verbal ability" (Coxon, 2012, p. 294), and were therefore not chosen for gifted and talented programs or even talent searches. The Project Talent finding implies that a group of potentially gifted students in STEM are missed because the important facet of their talent is not adequately assessed.

Checklists and nomination procedures often supplement standardized testing, but these more informal assessments have their drawbacks as well. For example, nominations can be biased and there is substantial between-school variation in gifted identification suggesting that there are reliability and validity concerns in the identification process (McBee, 2010). Combining these checklists with a battery of other types of assessments helps their usefulness, especially when considering students underrepresented in STEM areas (e.g., Lohman, 2009; Lohman & Lakin, 2008). Lohman suggested putting together evidence and looking at it in the context of the opportunities students have had to make determinations concerning the potential of a student to excel in STEM areas.

Further, even after taking into account high test scores or other evidence of aptitude, the potential for excelling in STEM fields requires interest and motivation in specific STEM fields (Subotnik, Olszewski-Kubilius, & Worell, 2009). The development, growth, and nurturance of STEM interest are things that need to be assessed in children. Hidi and Renninger (2006) pointed out that interest evolves in stages and that assessing interest has to be framed in the context of the child's individual development. This means that identifying and nurturing STEM interests requires looking at all aspects of a child's life, including his or her family context and the types of activities a child engages in outside of school. These informal types of assessments can provide important information that can help teachers develop children's STEM-related interests and aptitudes.

Another concern in identification of STEM talent that has not received a great deal of attention is that new technologies and ideas may change what is indicative of STEM aptitude and talent (Schleicher, 2010). For example, even very complex calculations can be carried out quickly and easily with high-speed computers. Hence, speed in carrying out more routine mathematical procedures may not be as important to identify and nurture as part of giftedness in STEM areas. Further, although STEM fields have become very specialized, there has also been a realization that individuals who can work across STEM disciplines can provide valuable contributions to solving the complex problems faced globally (Zhao, 2012). Hence, assessing the ability of individuals to think in "systems" terms across STEM disciplines may be key to helping societies solve problems in the future. The emphasis on crossing disciplinary boundaries can be seen in the development and assessment of integrated STEM curricula that involve problems that require integration of several STEM fields to solve (Honey, Pearson, & Schweingruber, 2014). The emphasis on interdisciplinary problems points to another important skill for those in STEM fields—the ability to collaborate and communicate with others.

Lastly, the issue of when to start searching for those with potential aptitude for STEM areas is an issue that has been debated (Subotnik et al., 2009) Most of the talent search programs emerge during the late elementary to middle school

... using appropriate identification protocols for assessment is extremely important, and identification of gifted STEM students should not be based on just one assessment. Although traditional tests capture some students with STEM talent, there is a need to consider assessments that are sensitive to differences in other areas, such as spatial ability to identify the potential of some students whose capability to excel in STEM might be otherwise missed.

years. Some advocate for earlier identification (e.g., Silverman, 1992). Because the development of expertise in a STEM area depends on building skills that are often specific to that domain, helping students begin to engage in experiences early on that facilitate the growth and development of those skills is important in nurturing students' performance. This is especially true for students who come from minority or socioeconomically disadvantaged backgrounds. Such students often do not have the access to resources to help facilitate the growth of sustained interest and learning.

The above discussion suggests consistencies with the National Association for Gifted Children (NAGC; 2008, 2013) that using appropriate identification protocols for assessment is extremely important, and identification of gifted STEM students should not be based on just one assessment. Although traditional tests capture some students with STEM talent, there is a need to consider assessments that are sensitive to differences in other areas, such as spatial ability to identify the potential of some students whose capability to excel in STEM might be otherwise missed. Likewise, NAGC supports early identification of students to help increase the probability of developing talents in a child.

Assessment to Nurture STEM Growth and Development in the Classroom

The issue of what classroom teachers can do to assess the classroom achievement of students who show talents in STEM areas can be informed by the literature on identification of students. The multitude of interventions designed for students who have talent makes simple advice on assessment difficult to give. Further, sometimes nurturing a student's talent may take place in out-of-school settings. Hence, the nurturing of STEM talent in the classroom requires that teachers assess not only the student's in-school progress, but what he or she does outside of school with others who potentially nurture that student's development.

Lohman (2009) provided a framework for conceptualizing the development of aptitude. Lohman noted that:

> The primary aptitudes for academic success are (1) prior knowledge and skill in a domain, (2) the ability to reason in the symbol systems used to communicate new knowledge in that domain, (3) interest in the domain, and (4) persistence in the type of learning environments offered for the attainment of expertise in the domain. (p. 971)

Hence, assessment of students is an ongoing process of discovery of what students do well, using dynamic kinds of assessments to see what they can do as their knowledge is extended further and consideration of how interested they are in a domain and seeing how much they are willing to work and stretch their knowledge. As Ericsson, Krampe, and Tesch-Römer (1993) pointed out, expertise requires deliberate practice that moves students beyond just practice to competence. Significant practice that goes beyond the everyday requirements is indicative of a student who has developed a strong interest and passion in a domain. Having an "aptitude perspective" means that the expertise domain of the field must be foremost when assessing the student (Lohman & Lakin, 2008).

STEM fields all have diverse skills as well as common underlying abilities. Further, they all have developmental trajectories where the nature of what "extraordinary" achievement means changes as students develop (Lohman, 2009). Although rapid calculation of basic arithmetic facts might be outstanding performance in elementary school, the ability to conceptualize a multistep applied problem might be more indicative of performance in high school. Spatial ability may be important to moving student toward a STEM area. Without the appropriate nurturing of that skill or recognition of its importance the development of expertise may not be possible.

Ziegler and Phillipson (2012) provided a systems perspective suggesting that exceptional performance arises from the complex interactions between the students' cognitive and motivational systems and the environmental contexts in which they develop. They noted that one-time snapshots of students fail to capture the complex trajectories that lead students to succeed in STEM fields. For teachers, this means using the prior assessments that previous teachers have provided to formatively plan curriculum and activities for a student and providing a rich picture of

> For teachers, this means using the prior assessments that previous teachers have provided to formatively plan curriculum and activities for a student and providing a rich picture of an exceptional student that can be used by those who will work with the student in the future.

an exceptional student that can be used by those who will work with the student in the future.

Considering Lohman's four areas for assessment, and one additional area that we feel is important to include (engagement outside of school in STEM-related activities), our discussion will turn to what classroom teachers can assess to support student development for successful STEM careers.

Prior Knowledge and Skills in a STEM Area

Both formal and informal measures of prior knowledge and skill are needed to understand the potential strengths or gifts that a child has to develop. Formal achievement measures are important to gauge the level of students' knowledge, but consideration needs to be made for a whole host of performance or alternative assessments. For example, to gain some access to what students do outside of school, a teacher might assign students a presentation about an out-of-school hobby or interest. Extended analysis of materials that the student brings for the presentation along with questioning to examine the depth of knowledge a student possesses is a useful technique to gain insight to what knowledge students bring with them that teachers can connect to in the classroom. Informal assessments that address students' backgrounds, STEM experiences, and family life can help better understand what motivates a child, what interests a child, and whether he or she has developed in-depth knowledge in some area that can be leveraged to build STEM knowledge.

The Ability to Reason in the Symbol Systems Used to Communicate New Knowledge

Most STEM fields have a core set of mathematical abilities that are associated with them. Fluency in mathematics and statistics allows a student to understand the models and methods that are part of research in STEM fields. Although there are a variety of descriptions of what underlies mathematical ability, there are some core characteristics of high levels of mathematical thinking. Krutetskii (1968/1976) pointed to several characteristics of exceptional mathematics ability that have found support in the mathematics education literature. Students who are more capable are able to recognize problem types and generalize exemplar problems more readily, the ability to more quickly automatize and shorten problem-solving processes, and they are able to think more flexibly (e.g., being able to reverse operations, construct negations, or test the limits) about mathematical constructs. Further, according to Krutetskii, mathematically talented students are more capable of viewing the world with a mathematical cast of mind, have more persistence, and strive not just for a solution, but for a more elegant solution. When it comes to assessing mathematical understanding in children with the potential for exceptional achievement, assessments of mathematical thinking often have to move beyond just answers and the current grade level. Assessments that examine students' recognition of problem types, assessments that show evidence of understanding the underlying basis for mathematical solutions, assessments that show student mastery of complex mathematical systems (e.g., matrix operations), and assessments that show evidence of flexible thinking about problems are required to uncover and facilitate talent in STEM areas.

> Students who are more capable are able to recognize problem types and generalize exemplar problems more readily, the ability to more quickly automatize and shorten problem-solving processes, and they are able to think more flexibly (e.g., being able to reverse operations, construct negations, or test the limits) about mathematical constructs . . .

Another important type of thinking and representation for those in STEM fields involves spatial systems. As noted earlier, students who have higher levels of spatial ability are sometimes missed in talent identification, and there are few assessments and little in the curriculum that nurtures spatial thinking. A National Research Council (2006) review of spatial thinking described a number of processes and systems. For example, understanding of coordinate systems, dimensionality of space, understanding the limits and biases among spatial systems, being able to use geographic information systems (GIS), being able to map spatial systems, and having the capability to move between different spatial representations, are just a few of the skills mentioned. Not only are these skills that involve understanding the ways in which we represent space available, but there are also abilities to work with more idiosyncratic mental models of phenomena that are often used by scientists to test out and generate theories (e.g., National Research Council, 2006, pp. 2–3). Our mental models can help us conceptualize the phenomenon of interest. They can also influence the way we represent things and carry out processes. For example, Stigler (1984) found that abacus training among Chinese children influenced their representation of numeric calculation. Unfortunately, despite the importance of the ability to think spatially and use mental models, assessments of these kinds of skills and systems are lacking (National Research Council, 2006).

Although spatial intelligence may have some heritable component, the development of spatial skills is experience dependent and can be trained (Nisbett et al., 2012; Uttal et al., 2013). For example, Coxon (2012) found that male students ages 9–14 who were identified as gifted showed improvement in some spatial skills after a LEGO robotics unit. Uttal et al. (2013) found in a meta-analysis that interventions to improve spatial skills were largely effective and suggested that increasing training in spatial skills could increase the number of individuals who are STEM ready in the workforce. An implication for the classroom teacher would be to help nurture students who show strengths in spatial thinking. An additional recommendation would be to make sure that students who are strong in verbal and quantitative areas develop spatial skills as well.

Interest in the Domain

As Ericsson (2014) noted, expert performance cannot be reduced down to general ability measures. Without interest and motivation, exceptional performance in STEM areas is not likely. The classroom teacher is in the position to assess and foster interest in students. Hidi and Renninger (2006) provided a framework for helping teachers to understand interest development. From that framework, teachers can look both formally and informally for evidence that a student has a "well-developed individual interest" (p. 112). Formally, teachers can look at interest inventories such as those that are part of the ACT tests, but as Hidi and Renninger pointed out, scores on the inventories do not tell us much about how or why students have interests in particular areas. Nor do the scores preclude the encouragement or development of other interests.

> Without interest and motivation, exceptional performance in STEM areas is not likely. The classroom teacher is in the position to assess and foster interest in students . . . . Teachers can work to engage interest by assessing where students appear to be situated in developing particular interests.

Informally, teachers can work to engage interest by assessing where students appear to be situated in developing particular interests.

Hidi and Renninger's model provides four phases in the development of interest. A teacher's sensitivity to the phase of interest development of a student may provide insight suggesting ways to further foster an interest. The first phase is **Triggered Situational Interest**. At this phase, a child may engage in some enriching activity (e.g., a science camp). That camp might lead to some initial interest in science, but unless other activities to foster that interest follow the camp, then that interest might fade. The second phase involves developing a **Maintained Situational Interest**, where there are contexts available to continue interest from an initial triggered event. A student gets involved with support or facilitation by a teacher or parent in additional camps or activities, additional reading, or might focus school projects in a particular area. The teacher's role would be to provide the student with opportunities to continue to pursue an area. Such opportunities may or may not facilitate the movement into the next phase. At this phase, the motivation to be engaged may be primarily external. As they move into the third phase of **Emerging Individual Interest**, students develop more enduring positive feelings and internally triggered interest in pursuing a particular domain with the support of others. At this phase, the teacher might notice a shift toward more internally generated interest that could move the child to a deeper interest. For instance, the child may ask a number of questions as he or she further delves into a domain. Hidi and Renninger also pointed out the importance of role models and supports that can help move the student into the next phase. In the fourth phase of **Well-Developed Individual Interest**, students are highly engaged and are likely to work beyond the given assignment and carry out more focused deliberate practice (Ericsson, 2014). All along the way from Triggered Situational to Well-Developed Individual Interest, it is the role of the teacher to assess what a child's needs are and to provide assignments and activities that address those needs. For example, at several phases, Hidi and Renninger noted that assignments that provide challenge and interaction are likely to move students to maintain a Well-Developed Interest.

> A teacher's sensitivity to the phase of interest development of a student may provide insight suggesting ways to further foster an interest.

Persistence in the Type of Learning Environments Offered for the Attainment of Expertise

Ericsson (2014) offered a compelling case that persistence in deliberate practice is the hallmark of those who obtain exceptional levels of performance in many domains. The persistence in performance of students who are identified as "gifted" in an area is often complicated by the meaning of giftedness to the student. Dweck (2009) noted a concern with the mindset created by the belief that ability is fixed and that effort means a lack of ability. If a student believes that high ability means never having to work hard, then when faced with challenge, such a student may interpret failure in challenging situations as indicative of a lack of ability. Dweck emphasized the need for children who are high performers on grade-level work to have the opportunity to work on more challenging tasks so that they can link success to effort and maintain such a mindset when faced with more challenging activities.

The self-determination theory (Deci & Ryan, 2000) focuses on opportunities for optimal challenge and autonomy in encouraging the intrinsic motivation to persist and work hard in school contexts. The development of classroom environments that encourage self-determination is an important issue for teachers to consider.

What are the implications of research theory on motivation and persistence for teachers? First, the teacher has to find ways to assess the attributions students make about success and failure. Formative assessment of how students react to feedback concerning their success and failure may be important to persistence in some students. Second, teachers need to ask whether they are providing gifted students with enough opportunities to engage in challenging work. Finally, teachers need to assess whether their classroom environment is supportive of self-determination. Although there are some measures to address these issues, informal conversations combined with examination of student reactions to challenge and failure can provide insight into the potential of students to persist.

> . . . teachers need to ask whether they are providing gifted students with enough opportunities to engage in challenging work.

Assessment of Out-of-School Activities

As noted in the earlier discussion about interest, the opportunities to take Triggered Situational Interest and turn it into a well-developed individual interest requires activities that are after school or out of school to help maintain and rejuvenate interest in an area. In STEM areas, robotic, engineering clubs, Science Olympiad, technology fairs, science fairs, and other activities can be indicative of the developing interests of a child with the potential to be exceptional in a STEM field. Out-of-school and after-school activities do not necessarily need to be formal either. As Bell (2009) pointed out, activities such as gardening, collecting rocks, working with computing, and other such activities often involve learning something about science or other STEM areas.

Informal assessment of the type and nature of such outside-of-school activities provides the teacher with places to connect formal learning to the students' extracurricular life. Additionally, it provides the teacher with opportunities to help facilitate the development of interest. These types of activities can be assessed through projects or other opportunities to share interests and hobbies with others at school. Conversations with parents and students focusing on the students' activities and interests outside of school can also provide insight about whether the student is engaging in activities that promote STEM learning and interests.

Performance Assessment of STEM activities

Although traditional tests can be used to assess student knowledge of STEM areas, the types of activities that are associated with more complex learning in STEM fields are often better assessed through projects, reports, and other activities that require problem solving, design of experiments, the creation of products, computer programs, databases, simulations, models, or other more authentic STEM activities. In addition, there is a movement toward developing integrated STEM activities that involve application of more than one STEM field (Honey et al., 2014). For example, engineering design has been introduced in K–12 education as a context for integrating applied math and science content (e.g., Harlan, Pruet, Van Haneghan, & Dean, 2014). These kinds of activities involve using performance assessments rather than standardized tests. The Common Core State Standards (CCSS) in mathematics as well as the Next Generation Science Standards (NGSS) both support these kinds of activities.

One approach to evaluating such assignments involves the development of rubrics. Most rubrics consist of a set of project or activity traits that are scored in terms of a handful of levels of performance that are described by a numeric scale or set of adjectives. The most effective rubrics are those that not only provide a score, but are descriptive about what that score means. Rubrics that are descriptive can be used for formative assessment purposes to improve performance. An additional consideration for developing rubrics in the context of working with gifted students is that it is important that the rubric have sufficient levels to provide for the potential for improvement in students who are operating at a higher level. Most projects and authentic activities can be judged not just by grade level standards, but also through the standards of expertise. Hence, there is a natural upward extension of the scoring criteria that can be used to provide a more challenging activity for talented students. Hence, they provide a meaningful way to make sure that student growth can be assessed as well as current levels of performance (McCoach, Rambo, & Welsh, 2012). Further, a clearly defined set of traits can help students to think more metacognitively about the task, an important skill to develop in students (VanTassel-Baska, 2014).

Many STEM projects or activities also involve collaboration, so rubrics and scoring systems that help gauge collaboration are important to developing skills in collaboration that will be important in adult life. Assessing group work involves making sure that the high-achieving students do not become the sole people doing the work. If they are doing the work themselves because they do not trust others or because the others want to take advantage of the students' abilities, then the collaborative process is not in place (Salomon & Globerson, 1989). Collaboration, communication, and other soft skills are important to STEM learning. STEM fields involve teams of individuals working together rather than lone individuals. Without that ability, exceptional performance levels may not be reached.

Portfolio assessments are beneficial in assessing developmental progress of talented students. Portfolios are a collection of different types of work that students produce throughout the school year. Johnsen (2008) noted that portfolios "sample a wide range of student work within a given domain, allow for student differences, and focus on growth over time" (p. 230). Although beneficial for all students, portfolios are particularly appropriate for gifted students in that teachers can craft learning activities to target STEM domains to show not only current achievement, but STEM aptitude as well.

Johnsen (2008) described four types of portfolio assessments: the Everything (or Developmental) Portfolio, the Product Portfolio, the Goal-Based Portfolio, and the Exemplary (or Showcase) Portfolio. What differentiates these separate portfolios is not only the content, but also the reason for creating the portfolio. For example, the Everything Portfolio is used when a teacher wants to evaluate students on a long timeframe, so examples of work from both the beginning and end of a school year or semester will be included. The Product Portfolio is used to measure a student's skills by aligning criteria with a set of standards, such as the CCSS or NGSS, so the included work is chosen to match the standards. The Goal-Based Portfolio demonstrates mastery of a particular skill, and teachers choose work that exemplifies mastery of said skill. Finally, the Exemplary Portfolio, most germane to assessment of gifted students in STEM fields, allows the students more choice in learning, and also requires students to explain why they chose to include certain work, thereby allowing the teacher to see the student's metacognition at work.

Teachers must consider how they structure their plan for curriculum assessment when working with advanced students, such as targeting high-level skills and allowing them to work on these projects that require deeper thinking and application. As stated earlier, an Exemplary Portfolio is one example of an assessment that can do just that. By including different types of work that are guided by the teacher but chosen by the students, students have more control over how to do certain tasks. The Exemplary Portfolio allows teachers to see the critical thinking that students use to accomplish their work while also having the ability to examine *how* the students learned. By further requiring students to explain why certain finished products, or stages of a finished product, are included, teachers can also come to understand even more about students' aptitude for a certain STEM field.

Summary

In this chapter we talked about the multifaceted nature of assessment for identifying, nurturing, and maintaining the potential for success of students identified as academically talented in STEM fields. That multifaceted approach has been difficult to implement in face of the focus on the meeting of minimal standards with schools that leave children who are academically talented out of the picture. New standards in math and science (CCSS and NGSS) provide contexts for building curriculum and assessments that can support some of the practices described here. However, the only way to really help nurture talent is to examine the whole child over his or her school career (Ziegler & Phillipson, 2012). It involves assessing the contexts students live in, formatively assessing their work, assessing their interests, and also assessing their beliefs about learning in STEM areas. Without consideration of these factors, it is unlikely that we will close gaps in the STEM workforce.

Discussion Questions

1. Why is student potential in STEM domains overlooked? What can we do to better identify STEM strengths among students?
2. Explain some reasons why an Exemplary Portfolio is the best choice of portfolios to assess gifted STEM students. What would you have students include in it?
3. Develop or find a problem-based task that could be used to help assess spatial ability.
4. How do we enhance or use our current curriculum and assessment to identify students with high ability in the STEM domains?

References

Andersen, L. (2014). Visual-spatial ability: Important in STEM, ignored in gifted education. *Roeper Review, 36*(2), 114–121.

Bell, P. (2009). Learning science in informal environments: People, places, and pursuits. Washington, DC: The National Academies Press.

Coxon, S. (2012). The malleability of spatial ability under treatment of a First LEGO League-Based robotics simulation. *Journal for the Education of the Gifted, 35*(3), 291–316.

Deci, E. L., & Ryan, R. M. (2000). The 'what' and 'why' of goal pursuits: Human needs and the self-determination of behavior. *Psychological Inquiry, 11,* 227–268.

Dweck. C. (2009). Self-theories and lessons for giftedness: A reflective conversation. In T. Balchin, B. Hymer, & D. J. Matthews (Eds.), *The Routledge international companion to gifted education* (pp. 308–316). New York, NY: Routledge/Taylor & Francis.

Ericsson, K. A. (2014). Why expert performance is special and cannot be extrapolated from studies of performance in the general population: A response to criticisms. *Intelligence, 45,* 81–103.

Ericsson, K. A., Krampe, R. T., & Tesch-Römer, C. (1993). The role of deliberate practice in the acquisition of expert performance. *Psychological Review, 100,* 363–406.

Harlan, J. M., Pruet, S., Van Haneghan, J. P., & Dean, M. (2014). *Using curriculum-integrated engineering modules to improve understanding of math and science content and STEM attitudes in middle grade students.* Retrieved from http://www.asee.org/public/conferences/32/papers/10284/view

Hidi, S., & Renninger, K. A. (2006). The four-phase model of interest development. *Educational Psychologist, 41*(2), 111–127.

Honey, M., Pearson, G., & Schweingruber, H. (Eds.). (2014). *STEM integration in K–12 education: Status, prospects, and an agenda for research.* Washington, DC: The National Academies Press.

Johnsen, S. K. (2008). Portfolio assessment of gifted students. In J. L. VanTassel-Baska (Ed.), *Alternative assessments with gifted and talented students* (pp. 227–257). Waco, TX: Prufrock Press.

Keating, D. P. (Ed.). (1976). *Intellectual talent: Research and development.* Baltimore, MD: Johns Hopkins University Press.

Kell, H., Lubinski, D., Benbow, C. P. (2013). Who rises to the top? Early indicators. *Psychological Science, 24*(5), 648–659.

Krutetskii, V. A. (1976). *The psychology of mathematical abilities in school children.* Chicago, IL: University of Chicago Press. [Original work published 1968]

Lohman, D. F. (2009). *Identifying academically talented students: Some general principles, two specific procedures.* In L. Shavinina (Ed.), *Handbook of giftedness* (pp. 971–998). Amsterdam, The Netherlands: Elsevier.

Lohman D. F., & Lakin, J. (2008). Nonverbal test scores as one component of an identification system: Integrating ability, achievement, and teacher ratings. In J. L. VanTassel-Baska, (Ed.), *Alternative assessments with gifted and talented students* (pp. 41–66). Waco, TX: Prufrock Press.

Mann, R. L. (2006). Effective teaching strategies for gifted/learning-disabled students with spatial strengths. *Journal of Secondary Gifted Education.* 27, 112–121.

McBee, M. (2010). Examining the probability of identification for gifted programs for students in Georgia elementary schools: A multilevel path analysis study. *Gifted Child Quarterly, 54,* 283–297.

McClain, M. & Pfeiffer, S. (2012). Identification of gifted students in the United States today: A look at state definitions, policies and practices. *Journal of Applied School Psychology, 28*(1), 59–88.

McCoach, D. B., Rambo, K., & Welsh, M. (2013). Assessing the growth of gifted students. *Gifted Child Quarterly, 57*(1), 56–67.

Missett, T., & Brunner, M. (2013). The use of traditional assessment tools for identifying gifted students. In M. Callahan & H. Hertberg-Davis (Eds.), *Fundamentals of gifted education considering multiple perspectives.* (pp. 105–111) New York, NY: Routledge.

National Association of Gifted Children (2008). *Position statement: The role of assessments in the identification of gifted students.* Washington, DC: Author. Retrieved from: http://www.nagc.org/sites/default/files/Position%20Statement/Assessment%20Position%20Statement.pdf

National Association for Gifted Children. (2013). *NAGC-CEC Teacher Preparation Standards in Gifted and Talented Education.* Retrieved from: http://www.nagc.org/resources-publications/resources/national-standards-gifted-and-talented-education

National Research Council. (2006). *Learning to think spatially: GIS as a support system in the K–12 curriculum.* Washington, DC: The National Academies Press.

Newcombe, N. (2010). Picture this: Increasing math and science learning by improving spatial thinking. *American Educator, 34*(2), 29–43.

Nisbett, R. E., Aronson, J., Blair, C., Dickens, W., Flynn, J., Halpern, D. F., & Turkheimer, E. (2012). Intelligence: New findings and theoretical developments. *American Psychologist, 67,* 130–159.

President's Council of Advisors on Science and Technology. (2010, Sept.). *Prepare and inspire: K–12 science, technology, engineering and math (STEM) education for America's future.* Retrieved from http://www.whitehousegov/ostp/pcast

Salomon, G., & Globerson, T. (1989). When teams do not function the way they ought to. *International Journal of Educational Research, 13,* 89–99.

Schleicher, A. (2010). *The case for 21st-century learning.* Retrieved from http://www.oecd.org/general/thecasefor21st-centurylearning.htm

Shea, D. L., Lubinski, D., & Benbow, C. P. (2001). Importance of assessing spatial ability in intellectually talented young adolescents: A 20-year longitudinal study. *Journal of Educational Psychology, 93*(3), 604–614.

Silverman, L. K. (1992). The importance of early identification of the gifted. *Highly Gifted Children, 8*(1), 5, 16–17.

Stigler, J. W. (1984). Mental abacus: The effect of abacus training on Chinese children's mental calculation. *Cognitive Psychology, 16,* 145–176.

Subotnik, R. F., Olszewksi-Kubilius, P., & Worrell, F. C. (2011). Rethinking giftedness and gifted education: A proposed direction forward based on psychological science. *Psychological Science in the Public Interest, 1,* 3–54.

Uttal, D. H., Meadow, N. G., Tipton, E., Hand, L. L., Alden, A. R., Warren, C., & Newcombe, N. S. (2013). The malleability of spatial skills: A meta-analysis of training studies. *Psychological Bulletin, 139,* 352–402.

VanTassel-Baska, J. (Ed). (2014). Performance-based assessment: The road to authentic learning for the gifted. *Gifted Child Today, 37*(1), 41–47.

Webb, R. M., Lubinski, D., & Benbow, C. P. (2007). Spatial ability: A neglected dimension in talent searches for intellectually precocious youth. *Journal of Educational Psychology, 99,* 397–420.

Zhao, Yong (2012). *World class learners: Educating creative and entrepreneurial students.* SAGE Publications: Kindle Edition.

Ziegler, A. &. Phillipson, S. N. (2012). Towards a systemic theory of gifted education. *High Ability Studies, 23,* 3–30. DOI: 10.1080/13598139.2012.679085

CHAPTER 9

Infrastructure of Comprehensive STEM Programming for Advanced Learners

Bronwyn MacFarlane, Ph.D.

Developing outstanding talent in the areas of science, technology, engineering, and mathematics (STEM) has been widely lauded as a national educational and economic imperative (Kettler, Sayler, & Stukel, 2014). STEM is an educational acronym describing a learning focus upon content understandings across these four fields with an interdisciplinary and applied approach. Ideally, a STEM educational program should begin during elementary school and continue seamlessly through higher education and into a professional career path. A common STEM learning goal is to introduce students to the scientific method and how it can be applied to real-world applications. Through STEM education curricula, teachers and students focus upon the understandings, skills, thinking processes, and habits of professionals in the STEM fields. As students become introduced to and increasingly familiar with computational thinking, problem solving, and creative innovations, it is hoped that their interest will be piqued to further explore these four fields.

The specialized chapters throughout this text provide extraordinary insight into the details associated with each element related to STEM education. Two chapters give us clear insight into STEM schools operating in different states, and six chapters give us specific details and recommendations for handling

 DOI: 10.4324/9781003238218-10

STEM content curriculum. This chapter seeks to take the aerial view to consider planning the infrastructure design and review for programmatic, curricular, and instructional features of cohesive STEM educational services.

Planning and delivering the infrastructure of a comprehensive STEM program is instrumental in the talent development of advanced learners. Articulating comprehensive curricular sequencing across early childhood, elementary, middle, secondary, college, and career development will contribute to the talent development of students moving toward professional establishment in a STEM field.

Benefits of STEM Education and STEM Degrees/Careers

According to the American Institute for Research (2012), for every one unemployed person with a STEM degree, there are 2.4 jobs available. However, for non-STEM employment, there is only one job opening for every 4.4 unemployed persons. This is well illustrated by Figure 9.1, which schematically represents career demand for STEM degrees in Arkansas.

To address the concern for STEM education in general, a variety of STEM programs are in use at the secondary level among regular education settings (e.g., New Technology Network High School, Project Lead the Way, The EAST Core), and at the postsecondary level (e.g., UTeach). The New Technology Network High School model integrates STEM education and extensive project-based learning throughout the entire curriculum. Project Lead the Way offers rigorous and innovative project-based curriculum in pre-engineering in middle and high school grades or biomedical sciences in high school. EAST Core demonstration projects extend EAST-based principles (Environment and Spatial Technology) into mathematics and science classrooms, allowing hands-on, problem-based learning to be connected to the STEM curriculum. In undergraduate settings, EAST provides a STEM teacher-preparation model to recruit college students majoring in STEM disciplines to consider becoming teachers (e.g., UAteach at the University of Arkansas in Fayetteville, UALRTeach at the University of Arkansas at Little Rock, and UCA STEMTeach at the University of Central Arkansas).

There are many existing programs across the country that focus on STEM education, and the National Consortium of Secondary STEM Schools

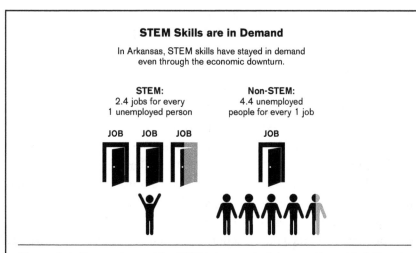

Figure 9.1. Career demand for STEM degrees in Arkansas. From *Vital Signs* by Change the Equation, 2014. Retrieved from http://vitalsigns.changetheequation. org/#ar-Arkansas-Demand. Copyright 2014 by Change the Equation. Reprinted with permission.

(NCSSS) provides a list of these schools, which can be found in the resources section at the end this chapter (Appendix 9.1). Many of these special schools were created with special legislation, like the Arkansas School for Mathematics, Sciences, and the Arts (ASMSA):

> ASMSA is one of sixteen public, residential high schools in the country specializing in the education of gifted and talented students who have an interest and aptitude for mathematics and science. ASMSA is unique in its mission and service to education. Created in 1991 by an act of the Arkansas Legislature, ASMSA has the distinction of being one of the nation's top secondary schools for superior students. All ASMSA instructors hold masters degrees and 27% have doctoral degrees. (ASMSA, 2011, para. 1)

Key Design Principles

The question of how to deliver a program that will develop and prepare young STEM innovators is a quandary facing many educators tasked with creating an optimal program. From the content curriculum chapters in this text it should be clear that high ability students need a STEM program that allows for acceleration, depth, complexity, flexibility, inquiry-based environments, authentic and interdisciplinary curricula, career exposure, mentorships, and career counseling.

The principles associated with gifted education can be examined in practice with the STEM programming operating at the Texas Academy of Mathematics and Science where students experience (1) acceleration in the content areas, (2) opportunities to work with advanced peer groups, (3) conducting authentic practice and research in STEM areas, and (4) additional challenge and motivation with academic contests (Kettler, Sayler, & Stukel, 2014). For educators tasked with designing, delivering, and leading STEM programming for advanced learners, it is important to learn from existing models such as the successful programs described throughout this text and take steps in order to build a robust infrastructure to meet the needs of advanced learners:

1. Establish clear goals for developing STEM talent,
2. Support and encourage acceleration and grouping for talent development,
3. Support authentic research,
4. Support academic competitions,
5. Remove the learning ceiling,
6. Provide meaningful and advanced professional development for teachers.

Educators involved with STEM programming design must consider the scope of key design principles for related decision making. For example, is there an option to build a special school building for STEM talent development, or is the program going to be embedded in an existing program? If educators have the special opportunity of building a new school, proposed plans must be generated about the facilities, location, operations, academics, growth, and communications dimensions.

In reviewing the current U.S. approach to serving adolescents who are talented and interested in STEM, Subotnik, Edmiston, and Rayhack (2007) found four models of state or privately funded programs that are designed to identify and develop STEM talent. They are:

1. Special schools, specifically at the secondary level emphasizing STEM subjects
2. Apprenticeship/laboratory programs such as afterschool or summer programs that provide hands-on opportunities to work in an authen-

tic STEM context and introductions by mentors into professional networks

3. Competitions that focus on STEM for middle and high school age students

4. Summer and afterschool courses at the middle and high school levels

Within their review of more than 100 STEM program websites, Subotnik, Edmiston, and Rayhack (2007) found that many STEM programs are missing important plans and policies regarding operations and learning outcomes. Likewise, in a review of the status of STEM High Schools across the nation, researchers at the University of Connecticut discovered many operational details with implications for practice (Gubbins et al., 2013). In a survey of high school students' perceptions of learning environments, fewer than half of the survey respondents (48%) claimed that they are challenged academically in "most" or "all" of their classes while a majority of students (63%) reported that they are not required to work hard in either "none" of their classes or only "1 or 2." STEM education is an interdisciplinary approach to leaning where rigorous academic concepts are coupled with real-world lessons. The following considerations and recommendations of programmatic, curricular, and instructional service features should be reviewed as educators plan and build upon STEM programs in local schools.

> Within their review of more than 100 STEM program websites, Subotnik, Edmiston, and Rayhack (2007) found that many STEM programs are missing important plans and policies regarding operations and learning outcomes.

Considerations for Planning Program Design

James J. Gallagher (1975) wrote that, "Failure to help the gifted child reach his potential is a societal tragedy, the extent of which is difficult to measure but which is surely great. How can we measure the sonata unwritten, the curative drug undiscovered, the absence of political insight? They are the difference between what we are and what we could be as a society" (p. 9). In his book, *Dumbing Down America* (2014), Jim Delisle wrote of his concern for and the importance of saving smart kids from educational structures that seem stacked against them and neither support nor extol excellence in practice, but in effect impede advanced students from moving forward quickly.

Just as gifted education programs should define which gifts and talents will be identified and cultivated in gifted services, so too, should STEM programs define STEM and how STEM-related topics will be operationalized for a program. However, no consensus exists thus far on definitions of STEM talent or on desirable outcomes for participants in STEM programs (Subotnik, Edmiston, & Rayhack, 2007). In addition to the core subject areas associated with STEM study, industrial clusters of interest to STEM curriculum include (a) agriculture, food, and environmental sciences; (b) biotechnology, bioengineering, and life sciences; (c) computer and information technology; and (d) nano-related and advanced materials and applications. These clusters of interest should also be considered in selecting STEM-related topics and local resources.

For school-based programs to chart a talent trajectory toward college and career, competitions such as *Math Counts* and *Intel Science Talent Search* are the only programs that claim to encourage STEM university majors and career pursuits and document student outcomes. To move toward this shared target for greater impact, goals and outcomes for participants must be established in program planning. Educational leaders must also plan regularly to measure the program's and students' outcomes.

> To move toward this shared target for greater impact, goals and outcomes for participants must be established in program planning. Educational leaders must also plan regularly to measure the program's and students' outcomes.

Feldhusen (1998a) defined a program model as a deliberately planned "system that facilitates interaction of gifted youth with curriculum to produce learning" (p. 211). Feldhusen also asserted that the identification process should find the students who match the program purpose (1998b). This wisdom of an optimal match across identification, program outcomes, and curriculum and instruction has been reiterated by multiple experts in multiple venues (VanTassel-Baska, 2003; Dixon, 2009; Renzulli, Gubbins, McMillan, Eckert, & Little, 2009; VanTassel-Baska & Little, 2011; Plucker & Callahan, 2014). Furthermore, these experts have reiterated the need for programs to be designed with a continuum of learning goals and plans to address the cognitive and socio-emotional needs of advanced learners. Learning goals should be integrated with researched best practices and relationships with established curricular programs, such as Advanced Placement and International Baccalaureate, as well as extracurricular activities such as Science Olympiad, Odyssey of the Mind, and Destination Imagination. Programs should also plan for students' career and college development opportunities and transitions, as well as consistent use of assessment data, technology, and professional development.

Details about the program should be easily accessible online for students, parents, and community members. STEM educational program websites should provide information about general selection requirements for admittance, specific STEM-related selection requirements, criteria demonstrating interest in STEM, desirable outcomes in general and specific to STEM for program participants, any gender-related priorities, and geographical boundaries defining participation (Subotnik, Edmiston, & Rayhack, 2007). In reviewing 117 programs for participation outcomes, only 34 suggested any STEM-related outcomes. By tracking and providing student and program outcome information, stakeholders will have a clear understanding and knowledge of the program's impact.

For program admissions, educators must consider what types of student information to collect, which may vary and could include standardized test scores in mathematics and science, teacher recommendations, STEM course completion requirements and prerequisites, STEM course grade levels, essays about STEM experience, or interviews. Of 78 special STEM schools reviewed, only 17 were found to require a stated interest in STEM as an admission criterion (Subotnik, Edmiston, & Rayhack, 2007). Delineating program features in advance and providing regular review will streamline the operations and public relations of the special program.

Curriculum Design Recommendations

Comprehensive curricular sequencing should be considered across early childhood, elementary, middle, secondary, college, and career development in STEM domains. Curriculum design and delivery for STEM programming can be guided by the principle that curriculum is a means to develop talents and research in gifted education. This suggests that curriculum based on in-depth conceptual understandings develops the talents and motivation of gifted students (VanTassel-Baska, 1987, 2000). Learner characteristics must be considered and a curriculum model for gifted learners should be selected to elegantly undergird all curricular work in the program. This section discusses these elements in detail as they relate to STEM curricula and what has been found to work effectively in supporting student STEM talent development.

Understanding Learner Characteristics

Primary age students have the most favorable attitudes toward science and teachers (Barrington & Hendricks, 1988). But science is taught more regularly as an integrated component in the primary school curriculum in countries outside the U.S., and very few STEM-related programs for middle school students reinforce deep existing interests that prepare individuals for careers in these fields (Subotnik, Edmiston, & Rayhack, 2007).

Student interest during the middle school years is malleable and can be reinforced with interaction and exposure to various career fields. When it comes to high-ability student interest in STEM, fewer than 10 states allow middle school students to take high school courses and zero allow them to take college courses (Subotnik, Edmiston, & Rayhack, 2007).

At the secondary level, academic work typically becomes more departmentalized. To support learner differences at the secondary level, Marshall, McGee, McLaren, and Veal (2011) described efforts by the Illinois Mathematics and Science Academy (IMSA) to help high-achieving economically disadvantaged youth. Through interviews, they collected information about what increased student motivation to attend the school and achieve, along with barriers in their environment that affected their education. Key factors that supported STEM learning specifically included the early use of multiple inquiry-based and problem-centered opportunities that foster mathematics and scientific reasoning and promote interpersonal and academic growth; a network of significant, caring, and trusting relationships; multiple opportunities to explore a wide range of interests and passions; and honoring the student's cultural heritage. Students also reported positive factors that contributed to their success, including attending a summer camp sponsored by the IMSA, mentorship by college students, family support, support from school personnel, and exposure to advanced technologies. Barriers included not having access to computers, peer pressure, the process of SAT registration, limited English at home, few supportive teachers, limited opportunities through public schools for different types of learning experiences, and limited school curriculum.

In examining the STEM pathway from K–12 to university, significant factors that affected talented students declaring STEM majors in college included measured ability, interest, self-efficacy, advanced academic coursework, gender, and set of different influences for selecting a STEM career (Heilbronner, 2011;

> Key factors that supported STEM learning specifically included the early use of multiple inquiry-based and problem-centered opportunities that foster mathematics and scientific reasoning . . . (Marshall et al., 2011).

Sadler, Sonnert, Hazari, & Tai, 2014). To provide a learner match between STEM talent development and curriculum, elementary science curriculum units developed in the Javits Project Clarion at the College of William and Mary Center for Gifted Education contained real-world problems facing today's society. The William and Mary curriculum units were based upon the Integrated Curriculum Model and designed to allow students to grapple with a complex issue, build a conceptual understanding of the issue as it related to larger ideas, and conduct scientific experiments. Empirical findings showed increases in students' understandings of macro concepts, science concepts, and scientific investigation and reasoning processes (VanTassel-Baska, Bracken, Stambaugh, & Feng, 2007). In the next section, the use of a curriculum model in a program for advanced learners is discussed.

The Integrated Curriculum Model

There are a variety of curriculum models for STEM programs to consider, such as the Schoolwide Enrichment Model, Autonomous Learner Model, Purdue Three-Stage Model, and the Integrated Curriculum Model. It is preferable for the selected model to have related empirical research (VanTassel-Baska & Brown, 2007) and the following discussion about the ICM provides an example of integrating a research-based curriculum model into advanced program design.

The Integrated Curriculum Model (ICM) was developed for high-ability learners to guide the process of differentiation. The model has three dimensions:

▷ an advanced content focus in core areas;
▷ high-level process and product work in critical thinking, problem solving, and research; and
▷ intra- and interdisciplinary concept development and understanding.

The ICM has been used as a basis to develop specific curriculum units in language arts, mathematics, science, and social studies that are aligned with state standards and differentiated for high-ability students. The ICM may be applied as a differentiation tool for remodeling curriculum in schools for high-ability students. Each learning activity or extension activity can be founded upon the structural design of the ICM. The three components of the ICM (see Figure 9.2) should be evident in each unit lesson (VanTassel-Baska, 2003; VanTassel-Baska & Little, 2011).

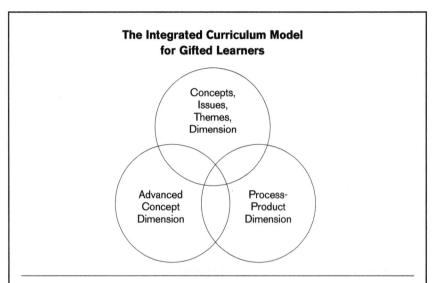

Figure 9.2. The Integrated Curriculum Model for Gifted Learners. From "The Development of Talent Through Curriculum," by J. VanTassel-Baska, 1995, *Roeper Review, 18*, p. 99. Copyright 1995 by The Board of Trustees of the Roeper School. Reprinted with permission.

By undergirding the subject-area content with a curriculum model such as the ICM, a concept-based curriculum framework will provide a robust foundation for planning the STEM curriculum and instructional delivery. The curricular plan should include: (a) learning goals, (b) a curricular framework, (c) an identified use of research-based curriculum models, and (d) identified instructional best practices. In applied practice, readers can refer to the science curriculum developed at the College of William and Mary as part of the Javits federal grant projects. Curriculum goals, for example, developed as part of William and Mary's Project Clarion include the ones outlined in Table 9.1.

VanTassel-Baska (2003) recommended using resources for high-ability students that are "interdisciplinary and idea-based," and materials that identify additional resources, providing "multiple options for reading or doing activities." High-ability and technically talented students should have learning experiences that provide them with opportunities to make connections with technology and across disciplines, apply models of teaching and learning to build their metacognitive and higher order reasoning skills, and build their confidence their STEM abilities (Robbins, 2010; Kettler, Sayler, & Stukel, 2014).

> The curricular plan should include: (a) learning goals, (b) a curricular framework, (c) an identified use of research-based curriculum models, and (d) identified instructional best practices.

TABLE 9.1

Project Clarion Science Curriculum Learning Goals

1. To develop selected basic concepts related to understanding the world of science
2. To develop selected macro concepts that unify understanding of basic concepts in science (i.e., systems, change, patterns, cause and effect)
3. To develop knowledge of selected content topics in science
4. To develop interrelated science process skills
5. To develop critical thinking skills
6. To develop creative thinking
7. To develop curiosity and interest in the world of science

From "Curriculum Goals" by College of William and Mary, n.d, retrieved from https://education.wm.edu/centers/cfge/research/completed/clarion/curriculum/index.php. Copyright 2015 by College of William and Mary. Reprinted with permission.

Another important element to be embedded in the infrastructure of a STEM program is that students should have opportunities to be mentored and work with an academic coach. Academic coaches at key developmental stages can be supportive for students pursuing the STEM fields to build upon their natural motivation and teach persistence and resilience for overcoming setbacks when challenged.

Recommendations for STEM Teacher Training and Program Evaluation

Today's educational climate is governed by accountability and standards, and assessment must be a component of every program and be considered and monitored in three ways: student learning, programmatic delivery, and personnel performance. Although testing can sometimes become a near focus over innovation, it does not have to be. For student assessment issues, Chapter 8 in this volume provides insights into examining student learning. For personnel and programmatic performance, the following ideas may be considered.

Training and Professional Development

Just as developmental learning is important for students, professional development plays a role in the STEM teaching force. Training, in-service, staff development, professional development, and professional learning are somewhat synonymous terms used to describe opportunities for teachers to improve their knowledge, skills, practices, and dispositions (Gubbins, 2014). For STEM educators who work with advanced learners, they are in need not only of advanced content understandings but also knowledge about learner characteristics and the application of differentiation features to curriculum and instruction. The literature surrounding the impact of high-quality teachers is robust and longstanding (Darling-Hammond, Wei, Andree, Richardson, & Orphanos, 2009). In designing and selecting professional development opportunities, alignment to professional teacher preparation and programming standards should be considered and are available from the National Association for Gifted Children (NAGC), as well as the associations for specific content areas (e.g., National Science Teachers Association, National Council of Teachers of Mathematics). Furthermore, specific training in properly using differentiated curriculum for high-ability learners is also recommended (VanTassel-Baska, Bracken, Stambaugh, & Feng, 2007; Robbins, 2010). For specific STEM curriculum details, readers should refer to the STEM content chapters presented in this volume for discipline-specific instructional strategies and relevant teacher preparation issues, such as confidence and self-efficacy for teaching STEM subjects.

Teacher Evaluation

It is imperative that teachers are highly trained in their subject area and in how to work with different populations of students in the same class, including the gifted and talented. A STEM teacher of gifted students must know and understand the specific subject content at a deeper level while also possessing strong facility of pedagogical skills. She must be attuned to her environment and the school functions around a set of customs and traditions (Cattani, 2002).

Unfortunately, too few U.S. teachers are qualified to teach STEM subjects, and without adequate teacher preparation and professional development, it will be difficult to improve student performance in those areas. Referencing quality teacher assessment instruments and tools specific to STEM education can provide important information to educators when examining personnel and programmatic performance. Appendix 9.1 provides a list of resources for additional information related to program review.

Program Evaluation

> A STEM teacher of gifted students must know and understand the specific subject content at a deeper level while also possessing strong facility of pedagogical skills.

Although a large amount of money is spent on STEM programs, there is a paucity of empirical evidence that STEM programs are effective in selecting and cultivating future STEM innovators (Subotnik, Edmiston, & Rayhack, 2007). STEM integration throughout the regular curriculum should be analyzed and considered. Subotnik, Edmiston, and Rayhack (2007) suggested that STEM program outcomes and data collection should include enrollment in postsecondary STEM programs, graduate satisfaction regarding preparation for STEM majors, achievement of graduates in STEM majors and/or careers, recognition for STEM-related activities and accomplishments, and awards for STEM related contexts received by program graduates. If multiple programs were to collect program outcome data, comparative studies could be conducted about program practices and what works in fostering STEM talent. By incorporating these elements to examine the definition, rationale, guiding principles, attributes, and related outcomes, educators can consider the strengths, weaknesses, and opportunities for improving the service delivery model of STEM educational content.

Conclusion

Providing integrated STEM educational experiences in an articulated delivery system demands that educators be skilled at working with curriculum, program design, and high-ability learners. Callahan (2001) suggested that provision of high-quality education to gifted learners demands a serious commitment of time, energy, administration, and funding; teacher expertise and deep understanding of academic content, thinking processes, instructional strategies, and student products; and a focus on student needs. Planning and revising the components related to delivering a sequenced STEM program is a critical aspect to the success of developing talent in the professional arenas described in this volume.

Discussion Questions

1. Identify and describe the planning process for reviewing and revising the STEM educational opportunities in the local school setting.
2. Discuss the curriculum models for meeting the learning needs of high-ability students, the components, and the implications for local application.
3. Discuss the professional development in STEM education needed in the local school setting.
4. Develop a plan for local programming to address STEM education and develop technical talent in specific STEM areas.

References

Barrington, B., & Hendricks, B. (1988). Attitudes toward science and science knowledge of intellectually gifted and average students in the third, seventh, and eleventh grades. *Journal of Research in Science Teaching, 25,* 679–687.

Change the Equation. (2014). Career demand for STEM degrees in Arkansas. Retrieved from http://vitalsigns.changetheequation.org/#ar-Arkansas-Demand

Cheuk, T. (2012). *Relationships and convergences found in Common Core State Standards in Mathematics, Common Core State Standards in ELA/Literacy, and A Framework for K–12 Science Education.* Arlington, VA: NSTA.

College of William and Mary. (n.d.). Curriculum goals. Retrieved from https://education.wm.edu/centers/cfge/research/completed/clarion/curriculum/index.php

Darling-Hammond, L., Wei, R., Andree, A., Richardson, N., & Orphanos, S. (2009). *Professional learning in the learning profession: A status report on teacher development in the United States and abroad.* Dallas, TX: National Staff Development Council.

Delisle, J. (2014). *Dumbing down America: The war on our nation's brightest young minds and what we can do to fight back.* Waco, TX: Prufrock Press.

Dixon, F. (Ed.). (2009). *Programs and services for gifted secondary students: A guide to recommended practices.* Waco, TX: Prufrock Press.

Feldhusen, J. (1998a). Programs and services at the elementary level. In J. VanTassel-Baska (Ed.), *Excellence in educating gifted and talented learners* (pp. 211–223). Denver, CO: Love Publishing.

Feldhusen, J. (1998b). Identification and assessment of talented learners. In J. VanTassel-Baska (Ed.), *Excellence in educating gifted and talented learners* (pp. 193–210). Denver, CO: Love Publishing.

Gallagher, J. J. (1975). *Teaching the gifted child* (2nd ed.). Boston, MA: Allyn & Bacon.

Gubbins, E. J., Villanueva, M., Gilson, C., Foreman, J., Bruce-David, M., Vahidi, S., . . . Tofel-Grehl, C. (2013). *Status of STEM High Schools and implications for practice.* Storrs: University of Connecticut, The National Research Center on the Gifted and Talented.

Gubbins, J. (2014). Professional development for novice and experienced teachers. In J. Plucker & C. Callahan (Eds.), *Critical issues and practices in gifted education: What the research says* (2nd ed., pp. 505–517). Waco, TX: Prufrock Press.

Heilbronner, N. N. (2011). Stepping onto the STEM pathway: Factors affecting talented students' declaration of STEM majors in college. *Journal for the Education of the Gifted, 34,* 876–899.

Kettler, T., Sayler, M., & Stukel, R. (2014). Gifted education at the Texas Academy of Mathematics and Science: A model for STEM talent development. *Tempo, 35*(1), 8–16.

Kettler, T., Sayler, M., & Stukel, R. (2014). Gifted education at the Texas Academy of Mathematics and Science: A model for STEM talent development. *Tempo, 35*(1), 8–17.

Marshall, S., McGee, G. W., McLaren, E., & Veal, C. C. (2011). Discovering and developing diverse STEM talent: Enabling academically talented urban youth to flourish. *Gifted Child Today, 34*(1), 16–23.

Plucker, J., & Callahan, C. (Eds.). (2014). *Critical issues and practices in gifted education: What the research says* (2nd ed.). Waco, TX: Prufrock Press.

Renzulli, J. S., Gubbins, J. E., McMillan, K. S., Eckert, R. D., & Little, C. A. (Eds.). (2009). *Systems and models for developing programs for the gifted and talented* (2nd ed.). Waco, TX: Prufrock Press.

Robbins, J. (2010). Adapting science curricula for high ability learners. In J. VanTassel-Baska & C. Little (Eds.), *Content-based curriculum for high-ability learners* (2nd ed., pp. 217–238). Waco, TX: Prufrock Press.

Sadler, P., Sonnert, G., Hazari, Z., & Tai, R. (2014). The Role of Advanced High School

Subotnik, R., Edmiston, A., & Rayhack, K. (2007). Developing national policies in STEM talent development: Obstacles and opportunities. In P. Csermely et al. (Eds.), *Science education: Models and networking of student research training under 21*. Amsterdam, The Netherlands: IOS Press.

VanTassel-Baska, J. (1987). A case for the teaching of Latin to the verbally able. *Roeper Review, 9,* 159–161.

VanTassel-Baska, J. (1995). The development of talent through curriculum. In *Roeper Review, 18*(2), 98–102.

VanTassel-Baska, J. (2000). The on-going dilemma of identification practices in gifted education. *The Communicator, 31,* 39–41.

VanTassel-Baska, J. (2003). *Curriculum planning and instructional design for gifted learners*. Denver, CO: Love Publishing.

VanTassel-Baska, J., Bracken, B., Stambaugh, T., & Feng, A. (2007). *Findings from Project Clarion*. Presentation to the United States Department of Education Expert Panel, Storrs, CT.

VanTassel-Baska, J., & Brown, E. (2007). Towards best practice: An analyses of the efficacy of curriculum models in gifted education. *Gifted Child Quarterly, 51,* (4), 342–258.

VanTassel-Baska, J., & Little, C. (Eds.). (2011). *Content based curriculum for high-ability learners*. Waco, TX: Prufrock Press.

Appendix 9.1: Additional Resources

Gifted STEM Educational Resources

▷ **National Consortium of STEM Schools (NCSSS):** http://ncsss. org/

▷ *Study of the Impact of Selective SMT High Schools: Reflections on Learners Gifted and Motivated in Science and Mathematics* **by Rena F. Subotnik, Robert H. Tai, and John Almarode:** http://sites. nationalacademies.org/cs/groups/dbassesite/documents/webpage/ dbasse_072643.pdf

▷ *Status of STEM High Schools and Implications for Practice:* https:// itunes.apple.com/us/book/status-stem-high-schoolsimplications/ id736858982?mt=11.

> ▷ **Jacob Javits Gifted and Talented Students Education Act and funded projects:** http://www.nagc.org/resources-publications/resources-university-professionals/jacob-javits-gifted-and-talented-students
> ▷ **Arkansas Project STEM Starters Javits Project:** http://ualr.edu/gifted/files/2011/11/Science_Fun_for_Everyone1.pdf
> ▷ **Colorado Project U-STARS Javits Project:** http://www.cde.state.co.us/gt/projectu-stars
> ▷ **The College of William and Mary Javits Project Clarion:** https://education.wm.edu/centers/cfge/research/completed/clarion/index.php
> ▷ **Differentiated science curriculum units from Prufrock Press:** http://www.prufrock.com/William-and-Mary-Units-C1185.aspx

Recommended Readings About STEM and Advanced Learners

Bruce-Davis, M. N., Gubbins, E., Gilson, C. M., Villanueva, M., Foreman, J. L., & Rubenstein, L. (2014). STEM high school administrators', teachers', and students' perceptions of curricular and instructional strategies and practices. *Journal Of Advanced Academics, 25*(3), 272–306.

Gubbins, E. J., Villanueva, M., Gilson, C., Foreman, J., Bruce-David, M., Vahidi, S., . . . Tofel-Grehl, C. (2013). *Status of STEM High Schools and implications for practice.* Storrs: University of Connecticut, The National Research Center on the Gifted and Talented.

Heilbronner, N. N. (2013). Raising future scientists: Identifying and developing a child's science talent, a guide for parents and teachers. *Gifted Child Today, 36*(2), 114–123.

Olszewski-Kubilius, P. (2010). Special schools and other options for gifted STEM students. *Roeper Review, 32*(1), 61–70.

Subotnik, R. F., Robert H., T., Almarode, J., & Crowe, E. (2013). What are the value-added contributions of selective secondary schools of mathematics, science and technology? Preliminary analyses from a U.S. National Research Study. *Talent Development & Excellence, 5*(1), 87–97.

Subotnik, R., Tai, R. H., Rickoff, R., & Almarode, J. (2010). Specialized public high schools of science, mathematics, and technology and the STEM pipeline: What do we know now and what will we know in 5 years?. *Roeper Review, 32*(1), 7–16.

General STEM Education Resources

- ▷ **EAST Initiative (Environmental and Spatial Technology):** http://www.eastinitiative.org/
- ▷ **New Technology Network High School:** http://www.newtechnetwork.org/schools/new-technology-high-school
- ▷ **Project Lead the Way:** https://www.pltw.org/
- ▷ **UTeach:** http://www.uteach-institute.org/

Appendix 9.2: National Consortium of Secondary STEM Schools List of Member Institutions by State

The National Consortium of Secondary STEM Schools (NCSSS) is an alliance of specialized high schools in the United States whose focus is to foster, support, and share the efforts of STEM-focused schools whose primary purpose is to attract and academically prepare students for leadership in mathematics, science, engineering, and technology. The Consortium supports unique professional development programs for STEM teachers and unique learning experiences for students.

The NCSSS was established in 1988 to provide a forum for member schools to exchange information and program ideas and to evolve alliances among them. As of June 2010, there are more than 90 institutional members, representing more than 40,000 students and 1,600 educators. These are joined by over 30 affiliate members, such as colleges, universities, summer programs, and corporations who share the goal of transforming STEM education. Below is a full list of NCSSS member institutions, by state.

Alabama
- ▷ Alabama School of Mathematics and Science
- ▷ Alabama School of Fine Arts—Russell Math & Science Center

Arkansas
- ▷ Arkansas School for Mathematics, Sciences and the Arts

California
▷ California Academy of Mathematics & Science
▷ Laurel Springs School

Connecticut
▷ Greater Hartford Academy of Mathematics and Science

Delaware
▷ The Charter School of Wilmington

District of Columbia
▷ McKinley Technology High School

Florida
▷ Center for Advanced Technologies
▷ Crooms Academy of Information Technology
▷ Mariner High School—Mathematics, Science and Technology Academy
▷ Middleton High School

Georgia
▷ Academy of Mathematics and Medical Science at South Cobb High School
▷ The Academy of Mathematics, Science & Technology at Kennesaw Mountain High School
▷ The Center for Advanced Studies in Science, Math, and Technology at Wheeler High School
▷ Rockdale Magnet School for Science and Technology
▷ Gwinnett School of Mathematics, Science, and Technology

Illinois
▷ Illinois Mathematics and Science Academy
▷ Proviso Mathematics and Science Academy
▷ Wheeling High School

Indiana
▷ Indiana Academy for Science, Mathematics, and Humanities

Kansas
- ▷ Kansas Academy of Math and Science

Kentucky
- ▷ Gatton Academy of Mathematics and Science in Kentucky

Louisiana
- ▷ Louisiana School for Math, Science, and the Arts
- ▷ Patrick F. Taylor Science and Technology Academy

Maine
- ▷ Maine School of Science and Mathematics

Maryland
- ▷ Anne Arundel County Public Schools
- ▷ Baltimore Polytechnic Institute
- ▷ Montgomery Blair High School
- ▷ Bullis School
- ▷ Oxon Hill High School, Science and Technology Center
- ▷ Poolesville High School
- ▷ Eleanor Roosevelt High School Science and Technology Center
- ▷ Science and Mathematics Academy at Aberdeen High School

Massachusetts
- ▷ Massachusetts Academy of Math and Science at WPI

Michigan
- ▷ Battle Creek Area Mathematics and Science Center
- ▷ Berrien County Mathematics & Science Center
- ▷ Dearborn Center for Math, Science & Technology
- ▷ Kalamazoo Area Mathematics and Science Center
- ▷ Lakeshore High School Mathematics/Science Center
- ▷ Macomb Mathematics, Science & Technology Center
- ▷ Mecosta-Osceola Mathematics/Science/Technology Center
- ▷ Utica Center for Mathematics, Science, and Technology

Mississippi
- ▷ Mississippi School for Mathematics and Science

Missouri
- ▷ Missouri Academy of Science, Mathematics and Computing

New Jersey
- ▷ Academy for Information Technology
- ▷ Academy of Allied Health & Science
- ▷ Bergen County Academies
- ▷ Communications High School
- ▷ Dwight-Englewood School
- ▷ High Technology High School
- ▷ Marine Academy of Science and Technology
- ▷ Marine Academy of Technology and Environmental Sciences
- ▷ Morris County Academy for Mathematics, Science and Engineering
- ▷ Red Bank Regional High School
- ▷ Union County Magnet High School

New York
- ▷ The Bronx High School of Science
- ▷ High School for Math, Science and Engineering at City College
- ▷ Hunter College High School
- ▷ Millennium Brooklyn High School
- ▷ Stuyvesant High School
- ▷ Brooklyn Technical High School
- ▷ Queens High School for the Sciences at York College

North Carolina
- ▷ North Carolina School of Science and Mathematics

Ohio
- ▷ Hathaway Brown School

Oklahoma
- ▷ Oklahoma School of Science and Mathematics

Pennsylvania
- ▷ Downingtown STEM Academy
- ▷ Pittsburgh Science & Technology Academy

South Carolina
▷ Dutch Fork High School
▷ The South Carolina Governor's School for Science and Mathematics
▷ Spring Valley High School

Tennessee
▷ School for Science and Math at Vanderbilt

Texas
▷ The Academy of Science and Health Professions
▷ John Jay Science and Engineering Academy
▷ Liberal Arts and Science Academy High School of Austin
▷ Academy of Science and Technology

Utah
▷ Academy for Math, Engineering and Science
▷ Northern Utah Academy for Math, Engineering, and Science
▷ SUCCESS Academies at Dixie State University and Southern Utah University

Vermont
▷ Essex High School

Virginia
▷ Health Sciences Academy Bayside High School
▷ Central Virginia Governor's School for Science and Technology
▷ Chesapeake Bay Governor's School
▷ Thomas Jefferson High School for Science and Technology
▷ LCPS Academy of Science
▷ Roanoke Valley Governor's School for Science and Technology
▷ Shenandoah Valley Governor's School
▷ Southwest Virginia Governor's School for Science, Mathematics, and Technology
▷ The Math & Science High School at Clover Hill
▷ The Mathematics & Science Academy
▷ Virginia STEAM Academy

Washington
▷ Camas Math, Science and Technology Magnet School

PART II

Applications to School-Based Practice With Gifted Education

Connecting the Common Core State, Next Generation Science, and Gifted Programming Standards With STEM Curriculum for Advanced Learners

Alicia Cotabish, Ed.D.

With the national focus on the Common Core State Standards (CCSS) and the newly released Next Generation Science Standards (NGSS), it is imperative to address how teachers of the gifted can deliver the goods without compromising content, knowledge, and skills. Both sets of standards highlight the curriculum emphases needed for students to develop the skills and concepts required for the 21st century. The adoption of the CCSS and NGSS has significant implications for teachers, particularly teachers of the gifted. Although both sets of standards call for general education teachers to recognize and address student learning differences and to incorporate rigorous content and application of knowledge through higher order thinking skills, they bear no obvious connection to the field of gifted education. The nature of advanced work beyond the standards is vague at best, although there is discussion about accelerating coursework in the CCSS appendix materi-

> The nature of advanced work beyond the standards is vague at best . . . . If held strictly to the standards, both the CCSS and NGSS could actually limit learning. However, teachers of the gifted are skillfully trained in the art of delivering differentiated curriculum and instruction. It is not a skill to be taken lightly, and one that becomes a pivotal point of necessity when addressing the standards while meeting the needs of advanced learners.

 DOI: 10.4324/9781003238218-10

als and an acknowledgement of the needs of gifted learners in Appendix D of the NGSS. If held strictly to the standards, both the CCSS and NGSS could actually limit learning. However, teachers of the gifted are skillfully trained in the art of delivering differentiated curriculum and instruction. It is not a skill to be taken lightly, and one that becomes a pivotal point of necessity when addressing the standards while meeting the needs of advanced learners. Many, if not most, gifted educators are charged with teaching multiple grades that can span grades K–12. Even as skilled masters of our trade, this daunting task can be overwhelming with regard to addressing the CCSS, the NGSS, and the *2010 Pre-K–Grade 12 Gifted Programming Standards*. The articulation of multiple sets of standards calls for us to work smarter, not harder.

A Smart Approach to Integrating the CCSS, NGSS, and Gifted Programming Standards

Integrating language arts, literacy, mathematics, and science through a highly articulated instructional approach is one to be considered. To do so, a roadmap for meaningful planning that elevates learning in all subject areas to higher levels of passion, proficiency, and creativity for all learners is needed. With the 21st-century skills as a backdrop for curriculum planning, gifted educators may be pondering how to integrate the CCSS and the NGSS in gifted programming. Let's take a look at the relationships among the NGSS, CCSS, and the *2010 NAGC Pre-K–Grade 12 Gifted Programming Standards*. Figure 10.1 depicts the relationships and their student-centered expectations.

As you can see, there is common ground and overlap among the standards. Adams, Cotabish, and Ricci (2014) described the configuration of Figure 10.1 in the following statement:

> Practices and portraits were grouped to illustrate student-centered expectations. The midpoint of the graphic demonstrates the relationship across student-centered expectations and/or similar tenets among the four sets of standards. Furthermore, standard-specific student expectations that do not overlap across the four standards are listed in separate boxes. Please note that the graphic does not account for overlapping that may occur among two or three standards (e.g., the use

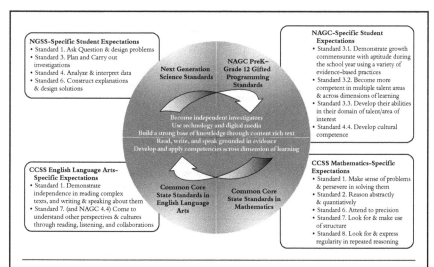

Figure 10.1. Relationships and convergences found in the NGSS, the Common Core State Standards in Mathematics, the Common Core State Standards in English Language Arts, and the NAGC Pre-K–Grade 12 Gifted Programming Standards. Adapted from *Relationships and Convergences Found in Common Core State Standards in Mathematics, Common Core State Standards in ELA/Literacy, and A Framework for K–12 Science Education*, by T. Cheuk, 2012, Arlington, VA: National Science Teachers Association. Copyright 2012 by NSTA. Adapted with permission.

of mathematical and computational thinking in both mathematics and science). (p. 5)

Using the standards and their similarities as a guide, curriculum units can be developed that address multiple standards across all of the disciplines. The author suggests examining the NGSS first when developing integrated units and planning instruction, because the NGSS have the benefit of being developed last; therefore, they are able to provide connections to specific linkages to the CCSS in mathematics and language arts (Achieve, Inc., 2014). Figure 10.2 depicts the layout of the NGSS standards and the location of the connections to the Common Core State Standards.

The NGSS connection boxes go beyond just listing the standard number of the connection to the CCSS. The boxes are inclusive of the title and specific connection to the CCSS. In addition, the NGSS include a helpful interactive website that allows the user to search and engage with the standards in two different ways—by Disciplinary Core Idea (DCI) or by Topical Arrangement. This

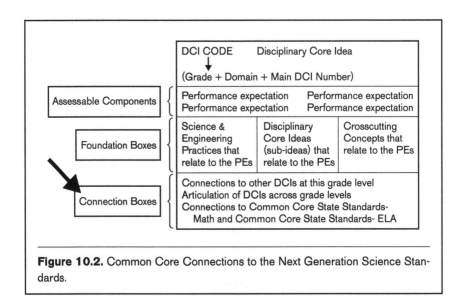

Figure 10.2. Common Core Connections to the Next Generation Science Standards.

tool is helpful in program planning and documentation. The website can be located at http://www.nextgenscience.org/next-generation-science-standards.

Cross-Disciplinary Content and Integrating Standards

As previously explained by Adams, Cotabish, and Ricci (2013), gifted educators can engage and motivate advanced learners by making cross-disciplinary connections. The following is an example of the strategy.

In the NGSS student expectation, *2-ESS-2: Construct an Argument Supported by Evidence for How Plants and Animals (including humans) Can Change the Environment to Meet Their Needs*, the linked CCSS mathematics standard requires students to model the explanation with mathematics, and the language arts standard requires students to write an explanatory text to examine the topic. Using mathematical and language arts concepts and skills, students can gain a deeper understanding of science content.

In Dimension 2 of the NGSS, crosscutting concepts are linked to each standard, including patterns, cause and effect, and systems. Some of these concepts are also found in the CCSS, making it easier to find points of intersection. Examining the standard as used above (*2-ESS-2*), students learning about how plants and animals can change the environment can explore the patterns and trends of such and use mathematics to model these patterns. Additionally, students can examine cause and effect influences of these parents and provide written or oral arguments with evidence to support their case.

Addressing the Inherent Limitations of the Standards

An important part of implementing the new standards is understanding the intent of them. Prior standards focused on disseminating understandings rather than student application and performances. Student expectations were vague and assessments were often misaligned with curriculum and instruction. The NGSS have avoided this difficulty by developing performance expectations that state what students should be able to do in order to demonstrate that they have met the standard, thus providing the clear and specific targets for curriculum, instruction, and assessment (Achieve, Inc., 2014). With this approach, the NGSS included clarification statements and assessment boundaries to guide discerning teachers in their curriculum and instruction planning. However, clarification statements and assessment boundaries can also impose unintended limitations to teaching and instruction. Although the NGSS acknowledge that this is not the intent, it could lead to unintended consequences that can directly affect advanced learners. To avoid this, the author suggests using the clarification statements and assessment boundaries as a point of departure for gifted learners—an avenue for acceleration and/or differentiation. For example, for the NGSS standard *K-P-S2-1: Plan and Conduct and Investigation to Compare the Effects of Different Strengths or Different Directions of Pushes and Pulls on the Motion of an Object*, the clarification statement reads: *Examples of pushes or pulls could include a string attached to an object being pulled, a person pushing an object, a person stopping a rolling ball, and two objects colliding and pushing on each other*. The accompanying assessment boundary states: *Assessment is limited to different relative strengths or different directions, but not both at the same time. Assessment does not include non-contact pushes or pulls such as those produced by magnets*. To add complexity for gifted students, the teacher could ask advanced learners to demonstrate pushes or pulls that consider relative strengths *and* different directions at the same time. In addition, the assessment could include noncontact pushes or pulls, such as those produced by magnets. Adding the element of prediction is another way to add complexity to the task. Although not formally stated in the standard, *prediction* is a stand-by tool that can always add complexity and relevance to student learning.

With regard to acceleration, vertical alignment/learning progressions are included in both the CCSS and NGSS. The graphic organization easily allows the gifted educator to see the pipeline of knowledge and skills that students engage in and provide a skeletal roadmap for accelerating students in content-specific disciplines who require such accommodations.

Examples of Existing Interdisciplinary Curriculum Guides for Advanced Learners

The Blueprints STEM Starters Series is a series of teacher curriculum guides focused on eminent scientists and inventors for whom exemplary children's biographies exist in trade book form (Robinson & Cotabish, 2005). In a recent STEM intervention study, teachers implemented STEM Blueprints to enrich gifted students' curriculum in conjunction with a problem-based learning science unit. The researchers reported increased student achievement in science concepts, content, and process skills when compared with students in a traditional science and gifted classrooms (Robinson, Dailey, Cotabish, Hughes, & Hall, 2014).

U–STARS Plus (http://www.fpg.unc.edu/node/4010), a previously funded U. S. Department of Education Jacob K. Javits project, produced "Science & Literature Connections" to explore scientific ideas within literacy instruction time using 32 popular children's books. "Science & Literature Connections" is organized around Bloom's taxonomy to support a range of thinking levels and to scaffold learning. By using these materials, a teacher can create a higher level thinking environment around literature connected with science, which motivates reluctant readers.

Seeds of Science/Roots of Reading (http://www.scienceandliteracy.org/) is another example of curriculum developed to integrate science and literacy. It was specifically designed for students in grades 2–5 and provides a focus on essential science understandings while building a full range of literacy skills. Using rigorous field testing, researchers found that students using Seeds of Science/Roots of Reading demonstrated increased student achievement in both literacy and science. Be mindful that this curriculum was not explicitly developed for advanced learners, and as with all curriculum, needs to be differentiated according to students' needs.

Conclusion

The adoption of the CCSS and NGSS requires gifted educators to navigate multiple standards as they plan differentiated curriculum and instruction for gifted learners. There are a number of strategies that teachers can use to implement and support the new standards for advanced learners at all stages of development. One aspect of fulfilling that purpose is to approach curriculum planning and instruction with an integrative approach. This strategy will allow teachers to address multiple standards at once and can be a smart, time-saving approach. Using the NGSS as the first stop for curriculum planning is particularly a smart move when attempting to make connections across the CCSS and NGSS. The interactive website and arrangement of the NGSS makes instructional planning less painful. It is also important to understand the inherent limitations of the CCSS and NGSS. Instead of viewing stated boundaries as limitations for gifted learners, the teacher of the gifted could use them as differentiated points of departure for advanced students. Lastly, there are wonderful differentiated curriculum resources that are readily available in our field and many more that are soon to come. It is important to note that although integrated materials are available, not all are differentiated for advanced learners. On this note, gifted educators will have to continue to be proactive advocates for advanced learners and recognize that the climate of education will continue to change as the larger field of education moves toward a student growth assessment model. With this in mind, it will be our responsibility to ensure that gifted and advanced learners are challenged and afforded the opportunity to intellectually develop, grow, and reach their potential.

Discussion Questions

1. What are some ways teachers of the gifted can provide the level of rigor and relevance within the NGSS and CCSS as they translate them into experiences for gifted learners?

2. With regard to standards implementation, how can teachers of the gifted provide creative and innovative opportunities that will nurture the thinking and problem solving of advanced learners?
3. In what ways can local educational reform (e.g., district policies and practices) in your district elevate learning to higher levels of passion, proficiency, and creativity for all learners?

References

Achieve, Inc. (2014). *Next Generation Science Standards*. Washington, DC: Author.

Adams. C., Cotabish, A., & Ricci, M. K. (2014). *Using the Next Generation Science Standards with gifted and advanced learners*. Waco, TX: Prufrock Press.

Cheuk, T. (2012). *Relationships and convergences found in Common Core State Standards in Mathematics, Common Core State Standards in ELA/ Literacy, and A Framework for K–12 Science Education*. Arlington, VA: NSTA. Retrieved from http://www.nsta.org/about/standardsupdate/ resources/VennDiagram–CommonCore–Framework.pdf

Robinson, A., & Cotabish, A. (2005). Biography and young gifted learners: Connecting to commercially available curriculum. *Understanding Our Gifted,* Winter 2005, *17*(2), 3–6.

Robinson, A., Dailey, D., Cotabish, A., Hughes, G., & Hall, T. (2014). STEM starters: An effective model for elementary teachers and students. In R. E. Yager (Ed.), *Exemplary science program series* (10th ed; pp. 1–18). [Monograph: National Science Teachers Association]. Arlington, VA: National Science Teachers Association.

CHAPTER 11

Biography Builds STEM Understanding for Talented Learners

Ann Robinson, Ph.D., Kristy A. Kidd, M.Ed., & Mary Christine Deitz, Ed.D.

Author's Note

This research was supported in part by a grant from the U. S. Department of Education, Jacob K. Javits Gifted and Talented Students Program. Correspondence concerning this manuscript should be addressed to Ann Robinson, Jodie Mahony Center for Gifted Education, College of Education and Health Professions, University of Arkansas at Little Rock, Little Rock, AR 72204. E-mail: aerobinson@ualr.edu.

Introduction

George Washington Carver, Marie Curie, Charles Darwin, Albert Einstein, Leonardo Fibonacci, Jane Goodall, Steve Jobs, Sally Ride, and Carl Sagan. Iconic figures in science, mathematics, and engineering provide talented learners with a portal to appreciating, understanding, and identifying with the STEM disciplines. Using biography to meet the affective and cognitive needs of talented

 DOI: 10.4324/9781003238218-13

students was most famously explored by Leta Hollingworth through her work with highly gifted children in selected New York City schools (Hollingworth, 1925, 1936). Developing a collection of biographies for the classroom with her own resources, Hollingworth determined that gifted children needed adult guidance in selecting biographies to read, but that they could manage their own discussions effectively (Hollingworth, 1925).

> Iconic figures in science, mathematics, and engineering provide talented learners with a portal to appreciating, understanding, and identifying with the STEM disciplines.

Fast-forward to the Common Core State Standards (CCSS) movement, and biography is specifically identified as a nonfiction reading choice to increase literacy achievement and to be used in service of science and social studies content (National Governors Association Center for Best Practices & Council of Chief State School Officers, 2010). Drill down into the Next Generation Science Standards (NGSS), and the integration of cross-cutting science concepts, science practices, and engineering design principles are well-represented in the new standards (NGSS Lead States, 2013). Given these two powerful educational standards initiatives, how does a nonfiction genre build STEM opportunities and achievement for children? In what ways does the use of biography, usually biographies of eminent adults, support STEM learning for talented students?

Eminence as a STEM Biography Seedbed

Biography and eminence are conceptually connected. The study of eminence has influenced gifted education from its early foundations in 19th-century psychology (Robinson, 2009, 2014). Victorian scholars in the emerging discipline of psychology and the established disciplines of biography and history were fascinated with the lives of creative, innovative, and talented "men of science." Galton conducted early studies of talent, including STEM talents, through the proxy of eminence that he defined as public acclaim accorded to individuals "decidedly well known to persons familiar with literary and scientific society" (Galton, 1874; VanTassel-Baska, 2014, p. 8). More recently, 20th-century scholars using biographical methods focused on the development of innovative and revolutionary ideas by eminent scientist Charles Darwin (Gruber, 1974), on cognitive case studies of creative people at their work (Wallace &

Gruber, 1989), on cross-case analyses of creative adults from a specific historical period (Gardner, 1993), and on eminent women in science (Filippelli & Walberg, 1997). Despite the initial Victorian creakiness and male-dominated perspective of the 19th-century fascination with eminence, examining a life for clues to the development of accomplishments and curiosity is an opportunity to convert the study of eminence into curricular opportunity for talented young children and adolescents (Robinson, 2009). The rationale for applying the study of eminent individuals in the STEM disciplines is captured by the earliest champion of using biography with gifted children, Leta Hollingworth (1936). She noted,

> The study of the life of civilized man leads inevitably to the study of biography. "What does the word 'pasteurized' mean on the milk bottle?" "Why is it called a volt?" "Why do they call it listerine?" These children need to know, and can learn, the relationship between civilization and the lives of significant persons "who really lived." (p. 89)

Literature on Using Biography to Teach STEM

The research on using biography to teach STEM can be characterized as action research reports or as qualitative case studies that focus on science biographies to illustrate aspects of science practices (Fairweather & Fairweather, 2010; Monhardt, 2005; Moore & Bintz, 2002), to encourage scientific thinking in children (Fingon & Fingon, 2009), to dispel stereotypical images of scientists and engineers (Hoh, 2009; Lovedahl & Bricker, 2006), or to investigate the use of STEM biographies in the gifted and talented classroom through teachers' perceptions (Deitz, 2012). A second strand of scholarship on biography in the classroom focuses on the thematic, topical, and textual analysis of children's biographies with respect to the perspectives conveyed by the biographers about scientists and science practices (Dagher & Ford, 2005), with attention to gender issues (Mori & Larson, 2006; Owens, 2009; Wilson, Jarrard, & Tippens, 2009). Although the literature on biography in the curriculum in gifted education dates back to the work of Leta Hollingworth and has been subsequently suggested by researchers in gifted education (Betts & Kercher, 1999; Robinson,

2006, 2009), the inclusion of biography as supplemental differentiated curriculum in STEM intervention models has been more recent (Robinson, Dailey, Hughes, & Cotabish, 2014; Robinson, Dailey, Cotabish, Hughes, & Hall, 2014).

Specifically, Fairweather and Fairweather (2010) reported that middle-level students' open-ended pre- and postassessment responses to a literacy-based unit on science biographies:

> demonstrated their awareness of the many different contexts in which scientific discoveries take place, especially the human aspects of the experience. Several students shared their newfound appreciation for the fact that scientists spent a long time, even years, working on the same problem. (p. 29)

In an evaluation on the use of contemporary biographies at the undergraduate collegiate level, Mori and Larson (2006) reported that participants found biographies to be the most memorable aspect of the course because students learned that scientists lead regular lives filled with the same challenges as everyone else. In other words, science was a career to which young women could aspire.

Finally, Deitz (2012) conducted a qualitative study of the implementation of science biographies by gifted education teachers participating in a Jacob K. Javits research and demonstration project, STEM Starters. Targeting children in grades 2–5, STEM Starters included problem-based science units and trade book biographies about scientists whose work was related to the concepts in the units. Data were collected through document analysis, interview and observation. The themes that emerged from the analysis indicated teachers found the use of STEM biography engaging for themselves and for their gifted students. Teachers were particularly positive about the use of portraiture provided in the *Blueprint for Biography* guides developed through the project and recommended portraiture as "the way in" to STEM biographies and to the science experiments included in the guides.

The second strand of research literature on the use of biography in STEM instruction is textual analysis of documents rather than classroom-based interventions. Textual researchers have examined biographies of scientists through the lens of specific criteria. For example, Dagher and Ford (2005) reviewed 12 children's biographies of scientists along three dimensions: how the biographers portray scientists, how biographers portray the nature and process of scientific knowledge, and how biographers portray the social processes of science. They noted that the use of children's literature has been strongly encouraged by science educators and hypothesized that one of the reasons for the repeated suggestions for trade books is the disappointment in science textbooks. Despite the frequent calls for biography in the science classroom, Dagher and Ford (2005) expressed concern for misrepresentation of science practices and for what Milne (1998) has termed the propagation of heroic scientist myths in children's biographies. They reported that biographies for younger readers did not communicate how scientists came by their knowledge, but tended to focus on early experiences of the scientists as children. They also noted that biographies of contemporary scientists rather than eminent historical figures such as Curie and Einstein tended to provide "richer descriptions of experimental science" (p. 377). Finally, Dagher and Ford (2005) expressed concern that biographies of scientists did not portray the full extent of the scientific practices, which included connecting evidence and theory and communicating results to the scientific community. They urged caution in the selection of biographies and recommended active teacher questioning to dispel possible misrepresentations. Nevertheless, they did acknowledge that biographies may "provide useful springboards for arousing student curiosity and interest in exploring the historical record" (p. 391).

> . . . biographies may "provide useful springboards for arousing student curiosity and interest in exploring the historical record" (Dagher & Ford, 2005).

Overall, the limited literature on the use of STEM biography indicates that biographies are frequently recommended and appear to be well received by students and teachers. They can misrepresent STEM content and the complexities of science as it is practiced in the real world, and thus require careful trade book selection and active instructional guidance by educators.

STEM Practices in General Education

One of the conceptual shifts in STEM education, particularly in science, is away from the linear steps of the scientific method to an emphasis on the interconnected nature of science as it is practiced and experienced in the real world (NGSS Lead States, 2013). Current STEM best practice suggests that teachers should direct the learner to the variety of creative yet distinctive methods used by various scientists as they work. Biographies provide the ideal opportunity to model what "real" science looks like. For example, Jane Goodall observed chimpanzees for extended periods of time, making detailed observations and

Current STEM best practice suggests that teachers should direct the learner to the variety of creative yet distinctive methods used by various scientists as they work. Biographies provide the ideal opportunity to model what "real" science looks like . . . . Through biographies of scientists, students begin to understand how scientific knowledge is developed and applied. Biographies also allow students to see that science has a greater purpose: to broaden our understanding of the world around us.

collecting empirical evidence, yet she did not follow the step-by-step experimental method still frequently taught in science classes today. In 1898, Marie Curie discovered that pitchblende emitted more radiation than expected, and subsequently dedicated 4 years of work to be able to prove that the element radium existed (Des Jardins, 2011; MacLeod, 2004). Through biographies of scientists, students begin to understand how scientific knowledge is developed and applied. Biographies also allow students to see that science has a greater purpose: to broaden our understanding of the world around us.

Regardless of the age of the student, proponents of connecting literacy with other disciplines argue that the use of picture books engages learners (Costello & Kolodziej, 2006). When applied to STEM biographies, the visual images such as portraits and photographs in picture biographies help students to see the scientist as a real person (Deitz, 2012). According to Ansberry and Morgan (2010), the format of picture books appeals to a more visually focused generation. The rationale for using biographies as an instructional resource in science is that the students develop a deeper understanding of the scientist's character as well as an understanding of his or her approach to the study of science. In addition, the use of multiple texts is valuable as a method for teaching inquiry and for students to gain a greater understanding of the essence of the scientist (Moore & Bintz, 2002).

STEM Practices in Gifted Education

STEM practices in gifted education have focused on curricular content acceleration, enriched exposure to disciplinary methodology, and an emphasis on real-world problems and scenarios (Jolly, 2009; Robinson, Drain, Kidd, & Meadows, 2014; VanTassel-Baska, Bass, Ries, Poland, & Avery, 1998). With respect to the focus of this chapter, the use of biography in STEM, best practice emphasizes the use of differentiated activities and talent development discussion questions to enrich curricular experiences for talented learners. As previously stated, the earliest champion of using biography in the classroom with gifted children was Leta Hollingworth. A proponent of curricular enrichment, Hollingworth (1925) essentially established a biography "club." After piloting a biography study for a 6-month period in the elementary grades, she provided students with the opportunity to organize their own discussion sessions but observed that without guidance, even very bright children were limited in their choices of the eminent figures whose biographies they wished to read. Consequently, Hollingworth began to build her own library of biographies. The biography reading club became such a popular instructional event that Hollingworth could not carve enough time out of the school day to satisfy the students' need for discussion. To address their need for further interaction and follow-up, Hollingworth instituted a classroom question box. Through what might be termed action research today, Hollingworth determined that biographical study with gifted students required at a minimum 60-minute biography club discussions each week for an academic year. In other words, minimal time allocation by the teacher and self-directed study by young gifted students resulted in an enduring educational practice in subsequent general and content-specific instructional and programmatic models in gifted education (Betts & Kercher, 1999; Robinson, 2009; Robinson & Schatz, 2002).

Trends in STEM Education That Align With Gifted Education Principles

The increasingly integrated nature of STEM education is a good match for the compressed curricular experiences recommended for talented learners. For example, one dimension of the NGSS includes integrated science and engineering practices. These practices direct the vision for STEM education toward understanding how scientific knowledge develops as well as the connections between science and engineering. The practices shift the focus of science experimentation to more general practices seen in all areas of science. These practices are summarized in Figure 11.1.

For example, one key science and engineering practice is the construction of models or explanations based on an investigation. The model or explanation is refined as empirical evidence is collected. Again, we can turn to the use of biographies to demonstrate what model construction looks like in the world of a scientist. As convincingly depicted in a children's biography by Birch (1996), Louis Pasteur was fascinated with the microbes he found in fermenting beet

Asking Questions (in science) and Defining Problems (in engineering)

Developing and Using Models

Planning and Carrying out Investigations

Analyzing and Interpreting Data

Using Mathematics and Computational Thinking

Constructing Explanations (in science) and Designing Solutions (in engineering)

Engaging in Argument from Evidence

Obtaining, Evaluating, and Communicating Information

Figure 11.1. Integrated science and engineering practices.

juice. He replicated the conditions in multiple models before discovering the ideal conditions in which these "black rods" would grow. Over the years, Pasteur continually modified his model in order to discover more about the ideal conditions in which this microbe would multiply. After reading *Pasteur's Fight Against Microbes*, students gain a deeper understanding of how models are used in science to develop questions and explanations, to generate data, and to analyze a system.

Unresolved Issues and Questions

Two unresolved questions in using biographies as curriculum enrichment in STEM education present themselves. First, how do teachers make choices about the inaccuracies or misperceptions found in some STEM biographies? Should such biographies be removed from consideration? Or is the inaccuracy or misperception an opportunity for discussion and investigation? Second, in what ways might educators incorporate biographies that do not meet the high bar of portraying the full arc of scientific endeavor as outlined by Dagher and Ford (2005)? Can a single child's biography, particularly for young readers, be expected to satisfy the biographer's demand for a story well told with the scientist's demand for conveying the full complexity of science and engineering discovery and design? Or can elements of STEM endeavors be more effectively learned and appreciated if multiple biographies are purposely selected and integrated across multiple grade levels to expose students to real-world practices? These currently unresolved questions about using biography to build STEM understanding are empirically testable.

Best Practices in Serving Gifted Learners Through STEM Biography

Best practice for incorporating STEM biographies into the curriculum for advanced learners derives from the confluence of understanding STEM talent development across the lifespan and by focusing on scientific concepts, science practices, and real-world problems. According to the NGSS Lead States (2013),

> From a practical standpoint, the Framework notes that engineering and technology provide opportunities for students to deepen their understanding of science by applying their developing scientific knowledge to the solution of practical problems. Both positions converge on the powerful idea that by integrating technology and engineering into the science curriculum, teachers can empower their students to use what they learn in their everyday lives. (p. 3, Appendix A)

The use of biographies in tandem with differentiated science or engineering curriculum can model for both educators and students the application of science concepts to practical problems. After reading and discussing *Always Inventing: A Photobiography of Alexander Graham Bell* (Matthews, 2006), students begin to understand how basic science concepts were only a piece of the puzzle. The science concepts were a stepping-stone to the design of an object that solved a problem. Bell's scientific understanding of human speech led to the design of a device that used electric current to carry sound through a wire.

Using *Blueprints for Biography* as a supplemental curriculum can tie the components of science knowledge, science process, and critical thinking skills together. Each Blueprint includes discussion questions and four types of enrichment activities: portrait analysis, persuasive writing prompts, primary source analysis, and point-of-view analysis (Robinson, 2006;

> Best practice for incorporating STEM biographies into the curriculum for advanced learners derives from the confluence of understanding STEM talent development across the lifespan and by focusing on scientific concepts, science practices, and real-world problems.

Robinson & Schatz, 2002). In addition, the Blueprints in the STEM Series includes an experiment related to the work of the scientist or engineer featured in the biography. Key features of the Blueprints that differentiate them for gifted learners are found in the discussion questions and in the interdisciplinary enrichment activities. All discussion questions are designed to elicit higher level thinking, but one type of question is most appropriate for gifted learners—the talent development question. Each Blueprint includes one or more discussion questions that focus on exploring, understanding, and accepting one's talents within the context of the STEM disciplines. For example, the following set of talent development questions taps students' understanding of key characteristics about scientists and engineers and the different roles individuals adopt (Hardy & Robinson, 2015):

> What does it mean to be an entrepreneur? What does it mean to be an inventor? Thomas Edison was both, but not all inventors are also entrepreneurs. Likewise, not all entrepreneurs are inventors. Create a Venn diagram listing qualities you would expect to find in an inventor, those you would expect to find in an entrepreneur, and those qualities you think inventors and entrepreneurs might share. Imagine yourself in the role of inventor and then imagine yourself in the role of entrepreneur. Which role appeals to you the most and why? (p. 19)

Using *Blueprints for Biography* as a supplemental curriculum can tie the components of science knowledge, science process, and critical thinking skills together. Each Blueprint includes discussion questions and four types of enrichment activities: portrait analysis, persuasive writing prompts, primary source analysis, and point-of-view analysis (Robinson, 2006; Robinson & Schatz, 2002). In addition, the Blueprints in the STEM Series includes an experiment related to the work of the scientist or engineer featured in the biography.

Biographies also allow students to see different approaches to solving a problem. Gorman and Carlson (1990) compared the inventive processes of Alexander Graham Bell and Thomas Edison and found that both Bell and Edison used abstract models, physical models, and problem-solving strategies. However, their use of these strategies was not the same. These observations led to the understanding that STEM inventors are unique, yet the generalized process of invention is the same. Thus, biographies fulfill the dual purpose of capturing a unique life, but also permit generalizations about key constructs in gifted education, such as creativity, invention, and innovation (Robinson, 2013).

Implications for Teacher Preparation

In addition to preparing teachers through adequate STEM content and content-specific pedagogical skills, incorporating biography as a vehicle to engage gifted learners in STEM requires that teachers understand biography as a specific type of nonfiction text. Biography has key features that distinguish it from its two closest literary cousins—autobiography and historical fiction. Teachers need the opportunity to develop their skills in selecting STEM biographies for talented learners. Figure 11.2 summarizes the recommended criteria.

In addition, the use of the four types of enrichment strategies incorporated in *Blueprints for Biography* implies that teachers are prepared to teach both portraiture and primary source analyses. While the use of primary source analysis aligns with the goal of developing the skills of a practicing professional frequently recommended for gifted students, these skills must be explicitly taught. In addition, portrait analysis is a specialized form of visual representation, and methods for teaching children about portraiture have been developed out of the informal learning institutions of art museums. Unless they have had content-specific pedagogical experiences in art history, most educators have not had the opportunity to learn the skills associated with portrait analysis. Thus, detailed instructional guides like the *Blueprints for Biography* are necessary for most teachers to feel comfortable with techniques used by docents and curatorial staff.

Summary

In summary, the use of biography to increase interest in and understanding of STEM disciplines has been recommended as a means to engage students, including talented students, in investigating the complexity of science practices, encouraging scientific thinking, and understanding the human face of STEM. To accomplish these instructional goals, talented students need opportunities for exposure to STEM biographies that portray scientists and engineers as role models engaged in the processes of discovery and problem solving, in persistent inquiry and design, and with a passion for their STEM talent area. Evidence suggests that talented students need guidance in selecting, reading, and discussing biographies to gain from the experience. To this end, educators should be supported with access to or with the resources to acquire STEM biography

Presents a compelling story

Provides accurate information about the individual and about STEM

Is sensitive to cultural and gender issues related to STEM underrepresentation

Incorporates and references primary sources

Includes examples of personal characteristics or processes relevant to STEM talent development

Figure 11.2. Criteria for Selecting STEM Biographies for Talented Learners.

collections they can use in the classroom and with professional development on differentiating biography instruction through talent development discussion questions and enrichment activities, such as portrait and primary source analyses. To build STEM understanding through biography, both talented students and their teachers need access to exemplary trade book biographies, time to pursue the rich lessons about the lives and work of scientists and engineers, and the opportunity to appreciate the affective and cognitive benefits of engaging with lives that inspire.

Discussion Questions

1. What are the key features of biography that distinguish it from other kinds of nonfiction reading? What criteria should be used in selecting STEM biographies for use with talented children and adolescents and why?

2. What key insights about the practice of science or cross-cutting concepts in science and engineering have been portrayed in children's biographies about scientists and engineers? An example would be the foci on observation as part of planning and carrying out investigations (practice) and on communicating information to the public (practice) explored in *The Watcher: Jane Goodall's Life with the Chimps* by Jeanette Winter (2011). What are some other examples from other children's STEM biographies?

Professional Learning Activities

1. Generate a list of scientists, mathematicians, and engineers who might be the subject of a children's biography. Browse library or book store collections. Search electronically for children's biographies on these individuals. Review the biographies with the criteria recommended earlier in this chapter.

2. Select a cross-cutting idea or a scientific practice from the Next Generation Science Standards. Locate a biography that enhances instruction about the cross-cutting idea or practice. Develop an alignment chart that summarizes the connection between the cross-cutting idea, the science practice, or the engineering design practice and the information or insights found in the biography.

3. Use the alignment chart (see Figure 11.1) to generate meaningful questions to guide the talented student as he or she reads so that a connection is made between the actions of a scientist or an engineer and the cross-cutting concept or practice portrayed in the biography. For example, how did the individual apply a cross-cutting concept such as Systems and Models or the practice such as Developing and Using Models to the discovery or innovation he or she made?

References

Ansberry, K., & Morgan, E. (2010). *Picture-perfect science lessons* (expanded 2nd edition using children's books to guide inquiry, 3–6). Arlington, VA: NSTA.

Betts, G. T., & Kercher, J. K. (1999). *Autonomous learner model: Optimizing ability.* Greeley, CO: ALPS Publishing.

Birch, B. (1996). *Pasteur's fight against microbes.* Hauppauge, NY: Barron's Educational Series.

Costello, B., & Kolodziej, N. (2006). A middle school teachers' guide for selecting picture books. *Middle School Journal, 38*(1), 27–33.

Dagher, A. R., & Ford, D. J. (2005). How are scientists portrayed in children's science biographies? *Science & Education, 14,* 377–393.

Deitz, M. C. (2012). *Gifted education teachers' perceptions on implementation of Blueprints for Biography: STEM Starters* (Unpublished doctoral dissertation). Little Rock: University of Arkansas at Little Rock.

Des Jardins, J. (2011). The passion of Madame Curie. *Smithsonian, 42*(6), 82–90.

Fairweather, E., & Fairweather, T. (2010). A method for understanding their method: Discovering scientific inquiry through biographies of famous scientists. *Science Scope, 33*(9), 23–30.

Filippelli, L. A., & Walberg, H. J. (1997). Childhood traits and conditions of eminent women scientists. *Gifted Child Quarterly, 41*, 95–103.

Fingon, J., & Fingon, S. (2009). What about Albert Einstein? Using biographies to promote students' scientific thinking. *Science Scope, 32*(7), 51–55.

Galton, F. (1874). *English men of science: Their nature and nurture.* Retrieved from http://www.mugu.com/galton/books/men-science/pdf/galton-men-science-1up.pdf

Gardner, H. (1993). *Creating minds: An anatomy of creativity seen through the lives of Freud, Einstein, Picasso, Stravinsky, Eliot, Graham, and Gandhi.* New York, NY: Basic Books.

Gorman, M. E., & Carlson, W. B. (1990). Interpreting invention as a cognitive process: The case of Alexander Graham Bell, Thomas Edison, and the telephone. *Science, Technology, & Human Values, 15*(2), 131–164.

Gruber, H. E. (1974). *Darwin on man: A psychological study of scientific creativity.* London, England: Wildwood House.

Hardy, B., & Robinson, A. (2015). *Blueprints for Biography: Young Thomas Edison.* Little Rock, AR: Jodie Mahony Center for Gifted Education.

Hoh, Y. K. (2009). Using biographies of outstanding women in bioengineering to dispel biology teachers misperceptions of engineers. *American Biology Teacher, 7*(8), 458–463.

Hollingworth, L. S. (1925). Introduction to biography for young children who test above 150 I.Q. *Teachers College Record, 2*, 277–287.

Hollingworth, L. S. (1936). The Terman classes at Public School 500. *Journal of Educational Sociology, 10*(2), 86–90.

Jolly, J. L. (2009). The National Education Defense Act, current STEM initiative, and the gifted. *Gifted Child Today. 32*(2), 50–53.

Lovedahl, A., & Bricker, P. (2006). Using biographies in science class. *Science and Children, 44*(3), 38–43.

MacLeod, E. (2004). *Marie Curie: A brilliant life.* Toronto, ON: Kids Can Press.

Matthews, T.L. (2006). *Always inventing: A photobiography of Alexander Graham Bell.* New York, NY: Random House.

Milne, C. (1998). Philosophically correct science stories? Examining the implications of heroic science stories for school science. *Journal of Research in Science Teaching, 35,* 175–178.

Monhardt, R. (2005). Reading and writing nonfiction with children: Using biographies to learn about science and scientists. *Science Scope, 28*(6), 16–19.

Moore, S., & Bintz, W. (2002). From Galileo to Snowflake Bentley: Using literature to teach inquiry in middle school science. *Science Scope, 26*(1), 10–14.

Mori, M., & Larson, S. (2006). Using biographies to illustrate the intrapersonal and interpersonal dynamics of science. *Journal of Undergraduate Neuroscience Education, 5*(1), A1–A5.

National Governors Association Center for Best Practices, & Council of Chief State School Officers. (2010). *Common Core State Standards for English language arts and literacy in history/social studies, science, and technical subjects.* Washington, DC: Authors.

NGSS Lead States. (2013). *Next Generation Science Standards: For states by states.* Washington, DC: The National Academies Press.

Owens, T. (2009). Going to school with Madame Curie and Mr. Einstein: Gender roles in children's science biographies. *Cultural Studies of Science Education, 4,* 929–943.

Robinson, A. (2006). Blueprints for biography: Differentiating the curriculum for talented readers. *Teaching for High Potential, Fall,* THP–7–8.

Robinson, A. (2009). Biography, eminence, and talent development: The lure of lives. In B. D. MacFarlane & T. Stambaugh (Eds.), *Leading change: The festschrift of Dr. Joyce VanTassel-Baska* (pp. 457–468). Waco, TX: Prufrock Press.

Robinson, A. (2013, April). *What can we learn from creative lives? Biography as method, data, and form.* Paper presented at the Annual Meeting of the American Educational Research Association (AERA), San Francisco, CA.

Robinson, A. (2014). Biography, history, and pioneering ideas: Illuminating lives. In A. Robinson & J. L. Jolly (Eds.), *A century of contributions to gifted education: Illuminating lives* (pp. 1–7). New York, NY: Routledge.

Robinson, A., Dailey, D., Cotabish, A., Hughes, G., & Hall, T. (2014). STEM Starters: An effective model for elementary teachers and students. In R. E. Yager & H. Brunkhorst (Eds.), *Exemplary STEM programs: Designs for success* (pp. 1–18). Arlington, VA: NSTA.

Robinson, A., Dailey, D., Hughes, G., & Cotabish, A. (2014). The effects of a science-focused STEM intervention on gifted elementary students' science knowledge and skills. *Journal of Advanced Academics, 25*(3), 189–213.

Robinson, A., Drain, L., Kidd, K. A., & Meadows, M. (2014). *Differentiating engineering is elementary curricula for talented learners in a summer enrichment program.* Manuscript in preparation.

Robinson, A., & Schatz, A. (2002). Biography for talented learners: Enriching the curriculum across the disciplines. *Gifted Education Communicator, 33*(3), 12–15.

VanTassel-Baska, J. (2014). Sir Francis Galton: The Victorian polymath. In A. Robinson and J. L. Jolly (Eds.), *A century of contributions to gifted education: Illuminating lives* (pp. 8–22). New York, NY: Routledge.

VanTassel-Baska, J. Bass, G., Ries, R., Poland, D., & Avery, L. (1998). A national study of science curriculum effectiveness with high ability students. *Gifted Child Quarterly, 42,* 200–211.

Wallace, D. B., & Gruber, H. E. (Eds.). (1989). *Creative people at work: Twelve cognitive case studies.* New York, NY: Oxford University Press.

Wilson, R. E., Jarrard, A. R., & Tippens, D. J. (2009). The gendering of Albert Einstein and Marie Curie in children's biographies: Some tensions. *Cultural Studies of Science Education, 4,* 945–950.

Winter, J. (2011). *The watcher: Jane Goodall's life with the chimps.* New York, NY: Random House.

CHAPTER 12

International STEM Education and the Integrated Role of Second Language Study

Bronwyn MacFarlane, Ph.D.

Although there has been an increase in American manufacturing productivity output since the turn of the 21st century, the percentage of manufacturing productivity in the U.S. economy has decreased, and in its place the services industry has boomed and grown faster. Relatedly, due to improved manufacturing processes, there has been a decrease in manufacturing employment and an increase in service-related employment. With these changes, the focus on STEM education has received heightened attention and support. Advocates for art education have expanded the discussion of STEM curriculum to purposefully integrate the arts into STEM learning experiences, per the corresponding chapter in this volume (Chapter 13). The rationale for integration of the arts in STEM studies includes a concern about producing technocrats with limited aesthetic appreciation for the arts and harkens to the longstanding wisdom in providing a liberal arts curriculum. At the same time, an influx of engineers to the U.S. from foreign countries has been increasing without the same kind of export of American engineers to foreign countries—despite the trend of more U.S. companies producing their goods overseas—and second language proficiency serves as the primary skill differential among domestic and international scientists and engineers.

 DOI: 10.4324/9781003238218-14

Since the Russians launched Sputnik 1, a call for the nation to cultivate a greater pool of fluent foreign language speakers has been consistently made in reports dating back to the 1960s (National Bureau of Economic Research, Arrow, & Capron, 1959; National Academy of Sciences, National Academy of Engineering, & Institute of Medicine, 2007; National Research Council, 2007). The National Research Council serves as an operating arm for both the National Academy of Science and the National Academy of Engineering, which focus on improving governmental decision making and public policy regarding science, engineering, technology, and health. In 1957, the National Defense Education Act was passed, which provided federal support for both foreign language and science education. In 1979, the President's Commission on Foreign Language and International Studies recommended foreign language requirements for all colleges and universities. By 1983, the College Board recommended expanding basic skills to include foreign language education for all students. In 1999, Secretary of Education Richard W. Riley delivered an annual Back-to-School Address, *Changing the American High School to Fit Modern Times*, in which he proposed second language study as the method to raise student performance on standards. The national education goals for 2000 also supported the need for second language learning with the statement, "By the year 2000 all American students will leave grades 4, 8, and 12 having demonstrated competency in challenging subject matter including . . . foreign language" (U.S. Department of Education, 1995, para. 3). But, the National Research Council's 2007 report about the progress of education in foreign languages in the U.S. noted that there was no apparent master plan and that the 14 U.S. Department of Education programs designed to strengthen education in foreign languages lacked the resources needed to keep pace with their mission. The report cited the need to develop an integrated approach to improving students' foreign language skills and expertise about other cultures and to improve upon an existing fragmented delivery approach (Zehr, 2007).

The 2006 National Security Language Initiative supported the teaching of languages considered critical to national security, such as Arabic, Chinese, and Farsi (Zehr, 2007), but the initiative did not receive funding and instead the Departments of Defense, Education, and State reorganized existing programs to carry out parts of the plan, resulting in the Education Foreign Language Assistance Program, which restructured grants for foreign language instruction at the K–12 level to go to the teaching of languages considered critical to national security rather than to more traditionally taught languages (Zehr, 2007).

In 2010, grant fund initiatives from the U.S. Department of Education Office of Postsecondary Education were announced to boost the numbers of trained teachers and, specifically, to support programs training teachers in STEM and foreign language instruction, including Arabic, Chinese, and Japanese (Brown, 2010). But, although the need for second language learning has continued to be recognized, the availability of second language learning opportunities has not continued to be offered. In fact, a counter shift began in schools during the early 21st century with a decline in local resources for second language course offerings. By 2007, the College Board reduced the number of foreign language Advanced Placement options by eliminating three course examinations, including French, Italian, and Latin literature.

To confound the issue, Forbes reported in 2012 that the percentage of public and private elementary schools offering foreign language instruction decreased from 31% in 1997 to 25% from in 2008 (Skorton & Altschuler, 2012). Foreign language instruction in public elementary schools dropped from 24% to 15%, with rural districts hit the hardest, and the percentage of all middle schools offering foreign language instruction decreased from 75% to 58%. Skilled teachers are also a premium in tackling the issue with a shortage of qualified foreign language teachers reported by about 25% of elementary schools and 30% of middle schools. Relatedly, by 2009–2010, only 50.7% of higher education institutions required foreign language study for a baccalaureate across degree programs, down from 67.5% in 1994–1995 (Skorton & Altschuler, 2012).

The case for the importance of studying a foreign language and beginning language study early has been made repeatedly by foreign language education associations (American Council on the Teaching of Foreign Language [ACTFL], 2006, 2008). As Stevens and Marsh (2005) pointed out:

> The learning of a foreign language exposes individuals to a range of new experiences. It touches not only upon social interaction, but also personal development and creative exploration, as well as intellectual and skills development. At its best, language learning opens up new worlds to learners within which self-discovery is a positive consequence. Individuals develop skills and acquire new dimensions of social interaction which, even at their simplest, open up new areas of communicative potential. (p. 113)

As direct foreign investment in the United States increases, the ability to speak fluently in a foreign language and understand another culture in diverse contexts will increase professional opportunities for Americans working with international companies. Even in Little Rock, AR, a traveler arriving at the local airport hears a welcome message at the baggage claim recorded in French, Spanish, and English. This chapter will provide a rationale that STEM programming should expand the curricular focus to include second language study with a range of language options including German, Spanish, Chinese, and Arabic. The use of second languages among high-performing STEM professionals, built from an educational base that provides for adequate second language study, will increase student opportunities for critical reasoning and skillsets to open an even wider range of opportunities in STEM fields.

Rationale for Integration: The Role of Second Language Education in Preparing for STEM Careers

The study of world languages has long been a valued component of a rich liberal arts education. Founders of American democracy were steeped in the study of classic Latin and Greek as well as various romance languages as they forged diplomatic alliances internationally and drew upon existing understandings to design a new country in a brave new world (MacFarlane, 2008). Understanding the communication and cultural competencies needed in a global world is fundamental for STEM students to be able to embark upon international STEM careers.

Learning a second language exposes children to new language systems, cultures, and helps them understand English better (VanTassel-Baska, 1987). Communication in a global world involves proficiency in language syntax, usage, and cultural literacy, which allow recognizing and participating in multilingual communities around the world (VanTassel-Baska, MacFarlane, & Baska, in press).

Although the study of a foreign language is recognized as an important component of a liberal arts education, several engineering programs do not have a second language requirement (University of Wisconsin-Madison, 2014; University of Arkansas at Little Rock, 2014, n.d.). More than half a century ago, the admissions office for the Massachusetts Institute of Technology (MIT) provided clear recommendations regarding foreign language study among prospective students in engineering, science, and technology to be in light of specific purposes (Olinger, 1946). At that point in time, MIT did not require proficiency in any foreign language as a prerequisite for admission, but rather urged that the choice to study a foreign language be made with a purposeful rationale for contributing to general education goals and thus making students more articulate and trained in expression, composition, and translation. Further, MIT recognized that studying an ancient or modern foreign language broadens students' cultural and intellectual horizons; cultivates knowledge of history, civilization, and philosophy; and develops skills for furthering international commercial, scientific, and social intercourse. Although the 1947 guidelines acknowledged that some practicing engineers may have little need of a second language, those who planned to work as research scientists should plan to have at least a reading knowledge of several foreign languages in order to understand the many scientific and technical reports, periodicals, abstracts, and other documents, including patents, not available in translation. Olinger (1946) wrote that when considering a purposeful rationale for foreign language study, French or German would be beneficial to engineering design work and Russian for interest in chemistry because these three languages were likely to be most important to a future scientist at that time. In looking ahead to predict global futures, there would be a greater importance for Americans in understanding Spanish, Portuguese, and Chinese. He also noted that Japanese, Scandinavian, and other languages may be of interest to scientists in certain special fields. Ultimately, MIT provided that in the absence of any such guiding principles, the selection of a specific language to study should depend on the student's best estimate of personal capabilities, interests, and probable future work, with an undergirding philosophy that "one should not own anything which he does not know to be useful or believe to be beautiful," so that the study of a second language should fulfill a definite use, existing or probable, or bring satisfaction and enjoyment.

> . . . the top universities known for attracting the most talented learners have incorporated second language study into baccalaureate graduation requirements.

Similarly, the University of Wisconsin engineering program of study (2014) did not require students to study a language but acknowledged that students live in a world in which a majority of people do not speak or read English, in which much of the knowledge that is disseminated may never appear in English, and in which students with sufficient study and preparation may be able to use a foreign language in their chosen discipline.

Moving forward over the last 70 years, some university engineering programs have remained somewhat uncommitted to requiring second language study, but the top universities known for attracting the most talented learners

As talented learners move into STEM careers in today's global world, they will find themselves working with diverse language speakers in settings ranging from health care, international business, security, military, and education.

have incorporated second language study into baccalaureate graduation requirements. In reviewing four of the consistently ranked top five engineering programs in the country, it is clear that students focused on STEM majors must have competency in a foreign language.

For admission to the University of Illinois at Urbana-Champaign, one of the top five engineering programs in the country, 2 years of successful high school foreign language courses are required for entrance. Undergraduates must be proficient in a chosen foreign language to the fourth-semester level either by test or course credits to graduate from the program of study. The College of Engineering offers area-focused minors including culture, recognizes language minors earned in the College of Liberal Arts and Sciences, and will preapprove courses at more than 32 top-rated engineering schools worldwide in more than 16 countries and consider others if requested. Over time, the MIT admission requirements changed and now require 2 years of high school foreign language from all incoming students. Furthermore, undergraduates are required to successfully complete three courses that are either in the same foreign language or are about the same world cultural area (e.g., West Africa). The Georgia Institute of Technology also requires 2 years of a foreign language for admission, and although it does not require the inclusion of a second language in an undergraduate program of study, it must include a "global perspectives" course for all students. Stanford University requires 3 years of successful foreign language study for admission. Stanford undergraduates must take one college year or test out of a foreign language; this requirement applies to students in all majors, including engineers.

To integrate second language learning into STEM programs, multiple elements must be considered, including learner characteristics for language acquisition, program design, curricular sequencing, and teacher skills and professional development.

Stanford offers a variety of study abroad programs, including 10 in which students take a semester abroad followed by an internship with a foreign company.

As talented learners move into STEM careers in today's global world, they will find themselves working with diverse language speakers in settings ranging from health care, international business, security, military, and education. The implications of these understandings point to a STEM program that should include multiple emphases, with a focus on STEM in an applied second language. Traditionally, STEM does not incorporate language literacy development within the acronym, and the limited focus on technical topics supports the rationale for theoretically making STEM plural (i.e., STEMS) for second language study. To integrate second language learning into STEM programs, multiple elements must be considered, including learner characteristics for language acquisition, program design, curricular sequencing, and teacher skills and professional development.

Second Language Learning and High-Ability Students

There is a correlation between language learning and students' ability to hypothesize in science. Bilingual children, given the same instruction by the same teacher to formulate scientific hypotheses, consistently outperformed monolingual children both in the quality of science hypotheses generated and in the syntactic complexity of the written language (Kessler & Quinn, 1980). Indeed, second language learning provides a natural connection for advanced learners, as it carries high interest. Like mathematics, it also provides another symbol system for providing challenge and complexity, two preferences that high-ability children demonstrate for learning new things (VanTassel-Baska, MacFarlane, & Baska, in press). Characteristics of advanced learners which match the second language educational experience include advanced vocabularies, wordplay, creative and curious interdisciplinary connections, opportunities for linguistic comparative analysis for adding depth and complexity, and divergent and convergent thinking processes (VanTassel-Baska, MacFarlane, & Baska, in press).

To explore gifted student attitudes toward modern foreign languages in the United Kingdom, 78 learners (ages 12–13) from two schools completed a questionnaire exploring their feelings about the prospect of being identified as gifted and talented in this subject area and their perceptions of the characteristics of highly able learners in modern foreign languages. Students were enthusiastic about the idea of being highly able in the subject area and expressed a fairly stereotypical view of the characteristics of being identified as highly able in the subject area (Graham, Macfadyen, & Richards, 2012), including possessing genetic high intelligence with fixed and innate ability factors.

Second language learning embedded in STEM studies offers an advanced interdisciplinary curriculum for talented learners to integrate understanding of the social customs, politics, arts, literature, and philosophy of a foreign culture through world language study. Thus it enhances cross-cultural competence and communication skills for the 21st century. As with mathematics, gifted learners can begin a second language early and accelerate that learning at a rate comfortable for them, often taking two second languages during their K–12 years (VanTassel-Baska, MacFarlane, & Baska, in press).

Student Scenarios Integrating STEM Studies With Second Language Learning

An integrated plan to study STEM and world languages must include early exposure to second language study. To consider integration of STEM and second language education, a variety of student scenarios at different levels can be considered when planning for talent development in global literacy with backward design. The following set of scenarios provides several case examples of student outcomes associated with integrating STEM with second language studies at different levels.

> Second language learning embedded in STEM studies offers an advanced interdisciplinary curriculum for talented learners to integrate understanding of the social customs, politics, arts, literature, and philosophy of a foreign culture through world language study.

Diana began language exposure to Chinese studies during elementary school and continued throughout secondary school and college. As a computer science major, Diana secured a part-time job with an electronics company, which led to an opportunity to travel to China with the company executives to scout potential manufacturing suppliers. The company executives were considering outsourcing to someone with Chinese translating skills, but because Diana had the desired skillset and had earned a solid reputation during her part-time job, she received the special opportunity to be included and work with the company executives.

Take, for example, Anya, who majored in engineering at the undergraduate and graduate levels. In high school, Anya had 4 years of German study, with her junior year spent abroad in a family home stay in Germany. As an undergraduate, Anya added a minor in German and went for a semester study abroad, taking 15–18 credits of engineering program coursework during the semester abroad, which were taught in German. Five courses were in engineering, and one course focused on the German language. Anya also conducted a research project in one of the professor's labs at the German university, which brought about invitations to return for doctoral research work. Anya's integrated experiences opened numerous opportunities with German engineering companies located in the United States, as well as opportunities with American engineering companies seeking to expand their global presence.

Cara studied Spanish for 2 years in junior high and all 4 years of high school. Spanish was chosen as a minor to complement her major area of study in the College of Arts and Sciences. During free time as an undergraduate, Cara volunteered her services to translate at a local shelter for women and children where Spanish-speaking skills were needed. After Cara continued graduate studies in medical school, she chose to practice in locations where she could serve and treat Spanish-speaking patients.

Eric began learning Arabic and Russian during his K–12 schooling years and continued to study the languages as a chemical engineering major. After Eric decided to specialize in petroleum engineering, his foreign language skills made him eligible for a plethora of opportunities to work with short-term and long-term projects across the Middle East and Russia.

Brian studied French for 4 years in high school and continued with coursework in college. Brian completed a summer study abroad program and stayed with a host family for part of the program and in university housing for the remainder. As a rising college senior, Brian applied for an international internship abroad with a major international company. Upon graduation, the internship experience contributed to the same company making Brian a full-time offer, which led to a mathematics career in the insurance industry.

Franklin began learning Japanese during his K–12 years when a Japanese exchange student stayed with his family. When Franklin chose to major in physics, he also chose to continue with Japanese as a minor area of study. Upon graduation, he received offers in the research and development departments from Japanese companies, including Sony and other high-tech manufacturers.

These student examples showcase the outcomes and relevant importance of developing second language literacy in concert with technical scientific disciplines. By shifting integrated world language study into STEM learning opportunities, educators will provide students with a better functioning curriculum to apply training in the science, technology, engineering, and mathematics fields.

STEM Programmatic Design and Instructional Delivery for Second Language Learning

In reviewing programmatic options for delivering specialized instruction in a second language for rapid language learners, it is important to note that not only has educating the gifted foreign language learner been marginalized in educational practice, the topic has also been marginalized in the research literature. Gifted education literature has expanded as more specialized research has been conducted about differentiated teaching in curriculum areas such as language arts, mathematics, science, social studies, and fine arts (VanTassel-Baska, Johnson, Hughes, & Boyce, 1996; VanTassel-Baska, Zuo, Avery, & Little, 2002; VanTassel-Baska, Bass, Ries, Poland, & Avery, 1998; Worley, 2006), but the literature on gifted and world language education is sparse (Robinson, Shore, & Enerson, 2007). There have been many studies conducted about foreign language learning in other countries, however. For example, there are case studies in China (Butler, 2014, 2015; Zhang & Yongbing, 2014), programmatic research in England and Germany (Andreyeva & Shaikhyzada, 2013; Busse & Walter, 2013; Graham, Macfadyen, & Richards, 2012; Yüksel, 2014), studies on European versus Asian foreign language education (Kobayashi, 2013), and foreign language teaching in South America (Barahona, 2014; Jorge, 2012).

We know that international educational findings do not provide an equal baseline for comparisons due to a range of differential student abilities measured and little focus on high-ability language learners. In American educational tradition, second language study has historically been offered at the secondary level as elective courses, attracting college-bound students. Over the years, the world language option offerings in the K–12 curriculum sequence have varied and enrollment has fluctuated (MacFarlane, 2012). The increasing dominance of English as a world language has made it hard for educational institutions to entice students to study languages other than English (Dornyei & Csizer, 2002). There is a conspicuous lack of studies exploring students' early foreign language learning experiences even though the importance of the initial year for further development has been well documented (Busse and Walter, 2013; Harvey, Drew, & Smith, 2006) and has been observed worldwide as the stage when student attrition is most likely to occur (Yorke & Longden, 2008). Research also suggests that the majority of second language learners wish they had begun a second language earlier than they did and felt it would be beneficial in procuring a job in the future (Clementi & Terrill, 2013). Indeed, the optimal time to begin children with second language learning is between the ages of 5–8.

Even though literature on the integration of STEM and second language study is sparse, integration is occurring in STEM programs as indicated by creative presentations at the American Council for the Teaching of Foreign Languages Annual Convention (Griffin, 2014). Information about planning and conducting a successful STEM German Immersion Day provided conference attendees with coordination details and program results about motivating students to study German by adding relevance with STEM topics and supporting students' professional growth and personal fulfillment. By using project-based units planned with the target language vocabulary, students were successfully engaged in using German and solving STEM problems. Furthermore, interdisciplinary connections between STEM and second language study are being made in special programs across the country, as exemplified by the Carol Martin Gatton Academy of Mathematics and Science at Western Kentucky University, the number one public high school in the country according to *Newsweek* and *The Daily Beast*, as described in Chapter 1 of this volume.

Curricular Design and Pacing Matters

To integrate second language learning with STEM education, teaching languages should not be treated as a separate curriculum but as a part of a curriculum unit's vocabulary and syntax. For gifted students a curriculum demanding more connections for STEM applications should be expected. Data about first-year university students enrolled in German courses in the UK revealed that despite students' increasing wish to become proficient in German, their effort to engage with language learning decreased over the course of the year. This changed occurred in conjunction with decreasing levels of intrinsic motivation and self-efficacy beliefs (Busse & Walter, 2013). Despite their high motivational levels at the beginning of the year, students did not have high initial self-efficacy beliefs. They reported problems of preparedness, insufficient opportunities for regular language practice with regard to speaking and listening practice, detachment and irrelevance of language exercises to the content of their degree courses, and indifferent relationships with language instructors who were contracted, as opposed to close relationships they experienced with other academic role models. To counteract decreasing motivation of modern foreign language students during their first year of university studies, the authors recommended that teaching languages should be integrated into the curriculum and not continue to be treated as a separate skill. This integration would make a vital contribution toward promoting advanced literacy skills, increased opportunities for use of the target language, and an increase in perceived value for language learning within the curriculum for students to sense.

Research in second language acquisition and pedagogy almost always yields findings that are subject to interpretation rather than giving conclusive evidence (Brown, 2002), and thus it may be clear that "enlightened" teachers would take an eclectic approach to language pedagogy with the use of dynamic classroom tasks and activities (MacFarlane, 2012). At least eight language teaching methods may be in practice today, including:

1. The Grammar-Translation Method
2. The Direct Method
3. The Audio-Lingual Method
4. The Silent Way
5. Suggestopedia
6. Community Language Learning
7. The Total Physical Response Method
8. The Communicative Approach

No comparative study has consistently demonstrated the superiority of one method over another for all teachers, all students, and all settings (College Board, 1986; Snow, 1994). How a particular method is manifested in a foreign language classroom depends heavily on the individual teacher's interpretation of the method's principles.

STEM programs integrating world language study should integrate target language vocabulary into the curriculum with the language instructor playing a role in the STEM training, and not treat language development as a separate skill with a separate teacher. ACTFL (2006) developed five national standards that are interdisciplinary in nature and identify the five C's for communication in a global world:

1. Communication: communicate in languages other than English
2. Cultures: gain knowledge and understanding of other cultures
3. Connections: connect with other disciplines and acquire information
4. Comparisons: gain insight into the nature of language and culture
5. Communities: participate in multilingual communities at home and around the world

STEM curriculum units designed with alignment to the ACTFL national standards will assist in developing global communicative skills about STEM content. The use of a gifted education curriculum model will provide the framework for planning differentiated curriculum units with focus on developing STEM global literacy. For example, a curriculum unit using the Integrated Curriculum Model (VanTassel-Baska, 1987) would incorporate the three dimensions of a an advanced concept, advanced content, and advanced thinking processes or products. To apply the Integrated Curriculum Model (ICM) with a curriculum unit focused upon electricity, curriculum planners could choose the advanced concept of Systems to undergird the unit and incorporate advanced content about electricity and associated vocabulary in the target language. Instructional learning activities to stimulate student thinking processes could be associated with building a circuit board using vocabulary in the second language and considering problems associated with global applications for electricity.

> STEM programs integrating world language study should integrate target language vocabulary into the curriculum with the language instructor playing a role in the STEM training

As STEM programs for talented learners continue to be designed, delivered, and evaluated, research should be collected to understand the dynamics associated with integrating world language learning. In a review of literature about content-based language instruction (CBI) in K–12 second language education programs in the United States, Tedick and Wesely (2015) isolated four broad themes characterizing the general foci of the extant research: (a) student outcomes, (b) classroom language use and development, (c) the hidden curriculum, and (d) teacher preparation and practice. The authors recommended future inquiry focus upon student diversity, the role of English in foreign language classrooms, teacher development, and achievement research. It will also be valuable to understand the impact of this shift in curriculum to measure the development of global communication literacy, creativity, and problem solving with global applications.

Instructional Strategies for Integrating Second Language Study Into STEM Education

Foreign language teaching methods are applied in a "local knowledge" context dependent upon particular institutions, classrooms, and learners. Shared constructivist viewpoints impacting the world language classroom include (a) students must be actively engaged in constructing personal knowledge and understanding, (b) new knowledge builds on previous knowledge and background schema plays a significant role in constructing meaning, (c) social interaction is critical in constructing knowledge and improving performance, (d) classroom tasks should mirror real-life tasks for students to practice application of their skills, (e) real-life tasks are grounded in context and purpose, and (f) assessment should reflect the complexity of integrating knowledge and skills into performance (Met, 1999).

In keeping with best practices in gifted education, strong curricula for world languages must engage students in higher level processes. World language teachers and curriculum writers should emphasize content-related issues while including attention to appropriate challenge, depth, complexity, and accelerative pacing. The curriculum should provide for both independent and interactive dynamic communications. The development of appropriate learner outcomes for gifted language students in world languages requires attention to the standards and best practices.

In order to infuse second language study into STEM subjects, the teachers of STEM, foreign languages, and gifted and talented must work together as a team to differentiate for talented learners and collaborate as coteachers. The curriculum writer also must consider an abundance of whole-class, group-work, and pair-work activities, as well as the plethora of "sophisticated" textbooks and resource materials exhibited at language teaching conferences regionally and nationally (Brown, 2002). Strategies that allow for more open-ended, interactive, and generative learning behavior are most beneficial to gifted learners (VanTassel-Baska, 2003). Although teachers may be left to formulate a sequence using these techniques, the use of a curriculum blueprint will assist in the "principled approach" to planning a vertical and horizontal sequence of developing gifted language learners (MacFarlane, 2010).

STEM and Foreign Language Teachers: An Integrated Team Approach

The skills developed in second language classrooms are needed for talented students who will pursue STEM careers and collaborate in providing services internationally. An integration of second language study into STEM educational programming for advanced learners can provide a contemporary platform of delivery in the 21st century. To integrate second language learning into advanced STEM programming in a meaningful way, all relevant personnel must be involved, including educators representing gifted, STEM, and foreign language programs; school leaders; and curriculum directors. Discussion must focus upon use of (a) existing foreign language program services, (b) inclusion of foreign language education in the articulated K–12 STEM curriculum, (c) relevant instructional issues with teacher training, and (d) creating policy related to providing services that meet the needs of advanced students (MacFarlane, 2012).

> An integration of second language study into STEM educational programming for advanced learners can provide a contemporary platform of delivery in the 21st century.

Although the case has been made for the value of a world language education component in a school curriculum and the value of second language study for gifted students, educational programmatic trends show concerns among isolated world language education programs (MacFarlane, 2008, 2009). The lack of instructional time necessary for acquiring a second language not only impacts student performance and self-efficacy, but teacher performance and individual expectations for students as well, thereby creating a cyclical effect as students enter world language teacher preparation programs and proceed to the classroom with their perceptions of past learning experiences (MacFarlane, 2009).

Three major qualities of exemplary secondary teachers of the gifted were isolated in a cross-cultural analysis:

1. strong content mastery,
2. a passionate personality dedicated to the teaching profession and students, and
3. a flexible and adventurous spirit in practicing instruction (VanTassel-Baska, MacFarlane, & Feng, 2006).

Yet, the ACTFL found that foreign language teachers need more training on best practices in teaching their subject area (2008). Compounding the ACTFL findings, another study found that Advanced Placement foreign language teachers needed targeted training in differentiated instruction and gifted education to serve gifted language learners in the AP foreign language classroom (MacFarlane, 2008). For advanced language learners, these findings help to explain and understand why only 3% of U.S. high school and college graduates achieve proficiency in a second language (Robinson, Shore, & Enerson, 2007).

Because gifted education and foreign language education programs are frequently marginalized in a comprehensive K–12 curriculum, professional development is recommended to increase awareness and appreciation of (a) second language learning as a critical component of an integrated curriculum, and (b) how teachers can meet the academic needs of advanced learners with sophisticated learning opportunities (MacFarlane, 2010, 2012).

Teacher training in the target world language and in differentiating instruction for advanced learners will be needed for an integrated STEM and second language education program. In a study of Advanced Placement world language teacher perceptions and instructional differentiation practices, teachers reported that they did not differentiate classroom instruction for gifted learners (MacFarlane, 2008). Professional development time should be designed to increase teacher self-efficacy in teaching global literacy with an integration of world languages and STEM education through specific curriculum planning sessions and content information. Area university programs in education, STEM, and foreign languages should collaborate to provide in-service workshops focused on building global literacy using STEM content.

Developing Fluent STEM Talent

In alignment with the top universities for training talent in the STEM fields, second language should be an integral component of a STEM curriculum for advanced learners. The incorporation of second language study into STEM educational programming should be designed with a conceptual base and sequence with the development of building content skills from novice to advanced higher linguistic operations. An integrated second language learning component within a STEM curriculum should be designed to encompass the educational preparation appropriate for talented learners to be ready for the global opportunities ahead.

Discussion Questions

1. Identify which curricular components can be reviewed and revised to integrate world language study into STEM curriculum.
2. Consider how an articulated curriculum can be sequenced to integrate STEM and second language learning through the K–12 program.
3. Opportunities for collaboration and professional development are necessary to integrate STEM and second language learning.

Brainstorm what these opportunities should include to meet local educational programming needs.

4. What are some methods to stimulate attention to the need for developing talented linguists and future communicators in the STEM professions?

References

American Council on the Teaching of Foreign Languages. (2006). *Standards for foreign language learning: Preparing for the 21st century.* Yonkers, NY: Author.

American Council on the Teaching of Foreign Languages. (2008). *Standards for foreign language learning.* Yonkers, NY: Author.

Andreyeva, O., & Shaikhyzada, Z. (2013). Communicative principle of teaching foreign language at non-linguistic specialties of the university. *European Researcher, 41*(2-2), 354–358.

Barahona, M. (2014). Preservice teachers' beliefs in the activity of learning to teach English in the Chilean context. *Cultural-Historical Psychology, 10*(2), 116–122.

Brown, E. A. (2010). Grants fund initiatives to boost number of STEM teachers. *Education Daily, 43*(125), 3.

Brown, H. D. (2002). English language teaching in the "post-method" era: Toward better diagnosis, treatment, and assessment. In J. Richards & W. Renandya (Eds.), *Methodology in language teaching: An anthology of current practice* (pp. 9–18). Cambridge: Cambridge University Press.

Busse, V. & Walter, C. (2013). Foreign language learning motivation in higher education: A longitudinal study of motivational changes and their causes. *Modern Language Journal, 97*(2), 435–456.

Butler, Y. G. (2014). Parental factors and early English education as a foreign language: A case study in Mainland China. *Research Papers In Education, 29*(4), 410–437.

Butler, Y. (2015). Parental factors in children's motivation for learning English: A case in China. *Research Papers in Education, 30*(2), 164–191.

Clementi, D. & Terrill, L. (2013). *The keys to planning for learning: Effective curriculum, unit, and lesson design.* American Council for Teaching of Foreign Languages. Retrieved from: http://www.actfl.org/publications/

books-and-brochures/the-keys-planning-learning#sthash.COEkEHkk. dpuf

College Board. (1986, Oct.). Speech delivered by Hanford, G. The SAT and Statewide Assessment. *Vital Speeches of the Day, 52*(24), 765.

Dornyei, Z., & Csizer, K. (2002). Some dynamics of language attitudes and motivation. Results of a longitudinal nationwide survey. *Applied Linguistics, 23,* 421–462.

Graham, S., Macfadyen, T., & Richards, B. (2012). Learners' perceptions of being identified as very able: Insights from modern foreign languages and physical education. *Journal of Curriculum Studies, 44*(3), 323–348.

Griffin, K. (2014). *Grow your German program.* Session presentation 083 at the American Council on the Teaching of Foreign Languages Annual Convention and World Languages Expo, San Antonio, TX. Retrieved from http://media.wix.com/ugd/0e2a77_4b81a9a53a298644 227ca5b45212e662.pdf

Harvey, L., Drew, S., & Smith, M. (2006). *The first year experience: A review of literature for the higher education academy.* York, United Kingdom: Higher Education Academy.

Jorge, M. (2012). Critical literacy, foreign language teaching and the education about race relations in Brazil. *Latin Americanist, 56*(4), 79–90.

Kessler, C., & Quinn, M. E. (1980). Positive effects of bilingualism on Science problem-solving abilities. In J. Alatis (Ed.), *Georgetown Universityround table on languages and linguistics* (pp. 295–308). Washington, DC: Georgetown University Press.

Kobayashi, Y. (2013). Europe versus Asia: Foreign language education other than English in Japan's higher education. *Higher Education, 66*(3), 269–281.

MacFarlane, B. (2008). *Advanced Placement world language teacher perceptions of high ability students and differentiated instruction* (Doctoral dissertation). Dissertations and Theses Database. (UMI Number 3319780)

MacFarlane, B. (2009). Global learning: Teaching world language to gifted learners. In B. MacFarlane & T. Stambaugh (Eds.), *Leading change in gifted education: The festschrift of Dr. Joyce VanTassel-Baska* (pp. 299–310). Waco, TX: Prufrock Press.

MacFarlane, B. (2010). Adapting world languages curricula for high ability language learners. In J. VanTassel-Baska & C. Little (Eds.), *Content-based curriculum for gifted and talented students* (2nd ed., pp. 283–302). Waco, TX: Prufrock Press.

MacFarlane, B. (2012). Perspective from the periphery: Teaching gifted foreign language students. *Tempo, 33*(2), 26–30.

Met, M. (1999). Research in foreign language curriculum. In G. Cawelti (Ed.), *Handbook of research on improving student achievement* (2nd ed., pp. 86–111). Arlington, VA: Educational Research Service.

National Academy of Sciences, National Academy of Engineering, & Institute of Medicine. (2007). *Rising above the gathering storm: Energizing and employing America for a brighter economic future.* Washington, DC: Authors.

National Bureau of Economic Research, Arrow, K .J., & Capron, W. M. (1959, May). Dynamic shortages and price rises: The engineer-scientist case. *Quarterly Journal of Economics*, 292–308.

National Research Council. (2007). *International Education and Foreign Languages: Keys to Securing America's Future.* Committee to Review the Title VI and Fulbright-Hays International Education Programs, M. e. O'Connell & J.L. Norwood, (Eds.). ISBN: 0-309-66939-1, 412 pages.

Olinger, H. C. (1946). Whiter foreign languages? *Modern Language Journal, 30*(7), 395–399.

Robinson, A., Shore, B., & Enerson, D. (2007). *Best practices in gifted education: An evidence-based guide.* Waco, TX: Prufrock Press.

Skorton, D., & Altschuler, G. (2012, Aug.). America's foreign language deficit. *Forbes.* Retrieved from: http://www.forbes.com/sites/collegeprose/2012/08/27/americas-foreign-language-deficit/print/staff_statements/index.htm

Snow, R. (1994). Abilities in academic tasks. In R. Sternberg & R. Wagner (Eds.), *Mind in context: Interactionist perspectives on human intelligence* (pp. 3–37). New York, NY: Cambridge University Press.

Stevens, A., & Marsh, D. (2005). Foreign language teaching within special needs education: learning from Europe-wide experience. *Support For Learning, 20*(3), 109–114.

Tedick, D. J., & Wesely, P. M. (2015). A review of research on content-based foreign/second language education in US K–12 contexts. *Language, Culture & Curriculum, 28*(1), 25–40.

University of Arkansas at Little Rock. (n.d.). Department of Systems Engineering: Study abroad programs. Retrieved from: http://ualr.edu/systemsengineering/programs/study-abroad-programs/

University of Arkansas at Little Rock. (2014). Department of Systems Engineering: Degree program. Retrieved from: http://ualr.edu/systemsengineering/programs/degree-program/

University of Wisconsin. (2014). College of Engineering: Answers to frequently asked questions for undergraduates. Retrieved from: http://www.engr.wisc.edu/future/ugradfaq.html

U.S. Department of Education. (1995). *National goals for education*. Retrieved from https://www2.ed.gov/pubs/EPTW/eptwgoal.html

VanTassel-Baska, J. (1987). A case for the teaching of Latin to the verbally talented. *Roeper Review, 9,* 159–161.

VanTassel-Baska, J. (2003). *Curriculum planning and instructional design for gifted learners*. Love Publishing: Denver, CO.

VanTassel-Baska, J., Bass, G. M., Ries, R. R., Poland, D. L., & Avery, L. D. (1998). A national pilot study of science curriculum effectiveness for high ability students. *Gifted Child Quarterly, 42,* 200–211.

VanTassel-Baska, J., Johnson, D. T., Hughes, C. E., & Boyce, L. N. (1996). A study of the language arts curriculum effectiveness with gifted learners. *Journal for the Education of the Gifted, 19,* 461–480.

VanTassel-Baska, J., MacFarlane, B., & Baska, A. (in press). *NAGC Seelct: Second language learning*. Washington, DC: National Association for Gifted Children.

VanTassel-Baska, J., MacFarlane, B., & Feng, A. (2006). A cross-cultural study of exemplary teaching: What do Singapore and the United States secondary gifted class teachers say? *Gifted Teacher International, 21*(2), 38–47.

VanTassel-Baska, J., Zuo, L., Avery, L. D., & Little, C. A. (2002). A curriculum study of gifted student learning in the language arts. *Gifted Child Quarterly, 46,* 30–44.

Worley, B. (2006). *Talent development in the performing arts: Teacher characteristics, behaviors, and classroom practices* (Unpublished doctoral dissertation). The College of William and Mary, Williamsburg, VA.

Yorke, M., & Longden, B. (2008). *The first year experience of higher education in the UK*. Retrieved from http://www.heacademy.ac.uk/assets/documents/research/FYE/FirstYearExperienceRevised.pdf

Yüksel, H. G. (2014). Becoming a teacher: tracing changes in pre-service English as a foreign language teachers' sense of efficacy. *South African Journal of Education, 34*(3), 1–8.

Zehr, M. (2007). NRC sees deficit in federal approach to foreign languages. *Education Week, 26*(31), 12.

Zhang, F. & Yongbing, L. (2014). A study of secondary school English teachers' beliefs in the context of curriculum reform in China. *Language Teaching Research, 18*(2), 187–204.

CHAPTER 13

Integrating the Arts and Creativity in STEM Education
Emerging Talent Using STEAM

Rachelle Miller, Ph.D.

Introduction

There is a demand for an increase in the number of individuals who are pursuing a STEM (science, technology, engineering, math) degree (National Science Board, 2006). Unfortunately, less than 40% of college freshman enrolled in STEM majors complete a STEM program of study (Higher Education Research Institute, 2010). These high attrition rates have led colleges and universities to take programmatic steps toward possible prevention of future drops in STEM discipline enrollment with mentoring and specialized interest groups. One intervention with potential to support the STEM pipeline found at the K–12 level is the integration of the arts into well-established STEM curriculum in gifted programs to provide students with another basis of engagement applying STEM content. This approach would allow STEM teachers to enhance gifted and talented students' learning by using the arts to create an integrated STEAM (science, technology, engineering, arts, math) curriculum. However, these efforts are best leveraged if embedded in curriculum earlier than college. STEM content in the classroom includes curriculum choices in K–12 envi-

 DOI: 10.4324/9781003238218-15

ronments focusing on improving student skills, preparing K–12 students for STEM degrees, and national competiveness in STEM fields.

Spatial and creative abilities are important for innovating in the STEM fields; however, these abilities rarely receive focused curriculum attention in schools (Coxon, 2012). Gifted and talented students exposed to the arts in connection to STEM content would have several advantages over students who have not experienced the arts integrated into technical fields, such as an improvement in long-term memory (Rinne, Gregory, Yarmolinskaya, & Hardiman, 2011) and an improvement in skills across content areas (Burnaford & Scripp, 2013; Catterall, Dumais, & Hampden-Thompson, 2012; DeBoer, Carman, & Lazzaro, 2010). The benefits of using a programmatic and curricular STEAM approach are twofold. First, this type of curriculum and environment helps to support and develop the convergent and divergent skills that the future innovators of our country must have in order to be competitive in the 21st century (Partnership for the 21st Century Skills, n.d.). Second, curricular STEAM experiences provide a bridge for gifted and talented students who do not typically gravitate toward STEM due to feelings of intimidation or lack of confidence with math or science to explore the fields with an artistic lens.

> Spatial and creative abilities are important for innovating in the STEM fields; however, these abilities rarely receive focused curriculum attention in schools (Coxon, 2012).

Incorporating the arts into a STEM curriculum allows students talented in other domains who may have traditionally eschewed math and science content areas to use different forms of expression to engage in, develop, and showcase math and science skills. These types of opportunities may even help nontraditional STEM students realize that they *are* good, and even strong, in math and science and realize that they can demonstrate their strengths in different ways. For example, Catterall and collegues (2012) found that incorporating the arts into STEM content both enhanced academic achievement and increased the participation of students from low socioeconomic backgrounds in STEM learning experiences.

The primary focus of STEAM is to integrate the arts and artistic design with science, engineering, and math across the K–12 educational range. Incorporating the arts into the STEM gifted and talented curriculum enhances literacy development as well as improves mathematical skills and understandings, observational science skills, and reasoning and critical response skills (DeBoer, et al., 2010). A well-developed STEAM gifted and talented curriculum will engage student learning and help students develop the skills related to "learning and innovation; information, media, and technology; and life and career skills;" needed to be successful in the 21st century (Partnership for the 21st Century Skills, n.d., p. 2). This chapter reviews empirical research and suggestions about gifted individuals producing innovation, the benefits of arts integration, teacher attitudes about arts integration, STEM integration of arts curriculum, and suggested recommendations for STEAM. The art integration examples relate to one or more of the following art subjects included in the National Standards for Arts Education (2014): dance, literary arts, media arts, music, theater, and visual arts.

> Incorporating the arts into the STEM gifted and talented curriculum enhances literacy development as well as improves mathematical skills and understandings, observational science skills, and reasoning and critical response skills (DeBoer, et al., 2010).

Gifted Individuals Producing Innovation

A society needs a group of innovators composed of not only scientists and engineers, but also of people knowledgeable in design, education, arts, music, and entertainment who interact with creative communities (Kerr & McKay, 2013). Innovation occurs when convergent thinkers, (those who focus on how to solve the problem) and divergent thinkers (those who explore many different solutions to a problem) join together to create novel ideas (Maeda, 2013). Convergent thinkers are more likely to be individuals who are math and science oriented, whereas divergent thinkers are more likely to be in the humanities and the arts (Furnham, Batey, Booth, Patel, & Lozinskaya, 2011).

It is possible that those who excel in technical fields do so because of an ability to think using both approaches. This may mean that eminent scientists and mathematicians are gifted as both convergent and divergent thinkers. Exemplars of eminent scientists may shed light on these abilities. For example, what did Luis Alvarez, Albert Einstein, and Hans von Euler-Chelpin all have in common? Not only were they all eminent scientists, but they also had profound interactions with the arts and science (Root-Bernstein & Root-Bernstein, 2013). Luis Alvarez was scientifically gifted, but he attended an Arts and Crafts school where he learned industrial drawing and woodworking. In 1968 he won the Nobel

> . . . what did Luis Alvarez, Albert Einstein, and Hans von Euler-Chelpin all have in common? Not only were they all eminent scientists, but they also had profound interactions with the arts and science (Root-Bernstein & Root-Bernstein, 2013).

Prize in Physics, and he believed that his success was due to his creative ability at building (Alvarez, 1987). Albert Einstein, who studied violin since the age of 6, attributed his scientific innovations to music: "The theory of relativity occurred to me by intuition, and music is the driving force behind this intuition . . . My new discovery is the result of music perception" (Suzuki, 1969, p. 90). Hans von Euler-Chelpin was a gifted Swedish biochemist who studied fine arts in college. His interest in painting and science led him to experiment with the color theory. In 1929, he won the Nobel Prize in Chemistry. These gifted innovators provide a few examples relevant to the many innovations created by gifted and talented individuals working in STEM fields and using divergent thinking skills and processes (Root-Bernstein & Root-Bernstein, 2013). John Maeda, former president of the Rhode Island School of Design and a strong advocate of STEAM, stated,

> Art helps you see things in a less constrained space. Our economy is built upon convergent thinkers, people that execute things, get them done. But artists and designers are divergent thinkers: they expand the horizon of possibilities. Superior innovation comes from bringing divergent (the artist and designers) and convergent (science and engineering) together. (Lamont, 2010, para. 4)

Piirto (2004, 2011) discussed various characteristics that describe highly creative people and the different processes that they experience. The "Five Core Attitudes" describe the five characteristics that creative people seem to possess:

- ▷ self-discipline toward their creative endeavor,
- ▷ open-mindedness toward experience,
- ▷ ability to take risks,
- ▷ trust of members in a group, and
- ▷ a tolerance for ambiguity.

The "General Practices for Creativity" that occur through the creative process include:

- ▷ working freely in solitude,
- ▷ completing rituals during or before creating their work,
- ▷ meditating formally or informally,
- ▷ exercising such as walking, and
- ▷ living a life of creativity.

The "Seven I's" are the various processes that creative people experience:

▷ being inspired,
▷ envisioning vivid images,
▷ imagining their work,
▷ having intuition,
▷ encountering insight, incubating through a period of rest, and
▷ improvising their creative work.

Unfortunately, with an emphasis on standardized assessments, secondary STEM gifted and talented students have been offered fewer opportunities to use creativity in their classrooms in recent years. Sousa and Pilecki (2013) suggested that secondary STEM students have been engaging in fewer creative activities because much of the current STEM curriculum content focuses upon memorization, minimal real-world experimentation, and convergent thinking. In addition, due to increased accountability via curriculum mandates, there may be too much content to cover in a STEM class. As a result, teachers may feel that they do not have the time to include arts-related activities in an already packed curriculum (Oreck, 2004).

Benefits of Arts Integration

Rinne et al., (2011) suggested that arts integration can improve students' long-term memory, particularly when educators incorporate activities that encourage students to:

▷ rehearse information and skills (i.e., repeat and drill),
▷ elaborate content through the use of artistic activities,
▷ generate information through artistic avenues instead of receiving it in written or oral form,
▷ physically perform and act out material,
▷ produce information orally,
▷ create meaning,
▷ express emotional responses to content, and
▷ present information in the form of pictures.

Arts integration enhances learning by providing students opportunities to work at the highest levels of Bloom's taxonomy—evaluating and creating. Arts integration can "introduce and create enthusiasm for a new unit of study; reinforce concepts already learned; and enrich current content by adding another layer of meaning" (Lynch, 2007, p. 34). By approaching STEM with an added arts component, STEAM, students are encouraged to engage and innovate with more imaginative playfulness in their learning (Johnson, 2014).

Academic achievement has been found to be positively associated with participation in arts-integrated programs. In partnership with the Chicago Public Schools, the Chicago Arts Partnership in Education (CAPE) delivered a 4-year project, Partnership in Arts Integration Research (PAIR), aimed to integrate the arts to improve student achievement (Burnaford & Scripp, 2013). The PAIR program included collaborations between elementary school teachers and teaching artists as well as professional development opportunities. Key findings and differences between experimental schools showed that arts integration assisted in elevating student achievement and closing the achievement gap, arts integration increased standardized tests, teacher involvement in arts integration positively influenced student achievement, and multiple assessments provided clear illustrations of student achievement.

In a comparison of student performance differences in arts-related and nonarts-related courses during the 2007–2008 and 2010–2011 academic years, academic performance results indicated that students who completed more arts credits showed higher academic achievement, fewer dropout rates, and scored higher on the SAT and the Florida Comprehensive Assessment Test (Kelly, 2012). A similar study examined the correlation between involvement in the arts and academic achievement of high school students from 2007–2010. Results revealed a positive relationship between students enrolled in arts classes and their academic achievement (Whisman & Hixman, 2012).

Teacher Attitudes About Arts Integration

Arts education has been shown to improve individual achievement in many studies, whether through traditional classes, extracurricular activities, or unique projects (Johnson, 2014). In order to create effective and appropriate arts integration professional development opportunities for teachers of the gifted and talented, it is necessary to understand teacher attitudes about incorporating the arts into established curricula. Oreck (2004) examined 423 teacher attitudes toward the arts using the *Teaching with the Arts Survey* (TWAS; Oreck, 2000).

Results indicated that teachers valued the arts and believed that the arts were important for student learning. However, these same teachers seldom integrated the arts into their curriculum because of lack of time and continued pressure to prepare students for standardized tests. Teachers also reported lack of confidence for incorporating the arts into their curriculum due to limited training in the arts.

In a follow-up to his 2004 study, Oreck (2006) explored the characteristics of six elementary teachers (i.e., four classroom teachers, one reading specialist, and one theater specialist) whose survey scores reported strong use of the arts and frequent arts integration of the arts into the core curriculum. Qualitative findings illustrated these teachers believed that the arts provided an avenue for differentiated instruction. These teachers cited professional development as a key factor that influenced their direct use of the arts in the classroom.

Studies examining successful teacher implementation of arts integration into classroom contexts found that teachers are more likely to use the arts when they receive support, such as by working with teaching artists (Andrews, 1999, 2006; Burnaford & Scripp, 2013; Garcia, 2003), having professional development opportunities (Andrews, 2008; Oreck, 2006; Patteson, 2002), or on-site art teachers (Andrews, 2010; Smithrim & Upitis, 2005).

Integrating the Arts and Creativity With STEM

Differentiating instructional strategies can assist in meeting the diverse needs of the gifted and talented (Tomlinson & Imbeau, 2010), and arts integration is one way of reaching a variety of students (Lynch, 2007). Moroye (2009) suggested that students may be more engaged when multiple senses are stimulated in the classroom, and that can be achieved with the arts. Unfortunately, there does not seem to be a standard protocol or approach for considering arts integration in STEM gifted and talented curriculum. Little research exists describing the integration of the arts into STEM content; however, those that do describe such efforts all yielded positive results (Barry, 2010; Moriwaki et al., 2012).

Moriwaki et al. (2012) described the Scrapyard Challenge Jr., a design workshop for children ages 4–11 that introduced the participants to basic concepts of electricity, conductivity, and mechanics. Workshop participants created interactive products of varying difficulty, depending on individual skill levels, using parameters of creativity, specified time frames, unpredictable materials, and electrical input/output boards. More than half of the participants successfully created a product. Overall, the workshop provided challenging, hands-on, real-world experiences that helped learners understand scientific and design concepts.

Barry (2010) examined the long-term effects that the Spencer Museum of Art (SMA) project had on teachers in various disciplines. For the purposes of this chapter, the cases describing STEM teachers integrating arts content are of most relevance. Qualitative findings of participant vignettes included two cases of STEM to STEAM scenarios. A middle school science teacher shared her experience teaching a unit about the human body with an art teacher. The middle school science teacher focused on the functions of the different bodily systems while the art teacher taught how DaVinci studied the human body in order to draw and sculpt. Overall, findings indicated that the SMA art project positively influenced teachers to integrate the visual arts into classroom curriculum. A high school mathematics teacher shared positive experience collaborating with the art teacher in cocreating a unit titled *Geoart*. In this unit, students created an area rug by applying various geometrical concepts, different types of tribal artwork, Islamic designs, as well as the artwork of other ethnic groups.

At the college level, Fantauzzacoffin, Rogers, and Bolter (2012) described an integrated art and engineering course offered at Georgia Institute of Technology. This college course consisted of art and engineering students from various fields. The goal of this course was for students to complete interactive projects using practices, insights, and perceptions from both the arts and engineering disciplines. Students were required to create projects in their field as well as design the process and criteria for their finished products. Positive feedback from participants supported the researchers' suggestion that creativity should be integrated by using project insight rather than by being solely problem based.

Suggestions for Integrating the Arts Into the STEM Gifted and Talented Curriculum

A well-developed STEAM curriculum integrates the arts with science, technology, engineering, and mathematics in which STEM curriculum and art standards should be connected seamlessly to provide rich learning experiences. Sousa and Pilecki (2013) offer various suggestions for incorporating the visual arts, theater, and creativity into the K–12 curriculum.

Grades K–4

▷ Create a word wall with vocabulary in the upcoming science unit. Provide opportunities for the students to use these words so they can become familiar with their meanings *before* teaching the unit. This will allow students to be more confident when learning new science content because they are already comfortable with related vocabulary.

▷ Integrate as many content areas as possible; create a short theatrical skit in the form of a mystery play incorporating writing exercises *and* showcasing understandings from a mathematics or science unit.

Grades 5–8

▷ When introducing a new science unit (e.g., energy), launch a contest about integrating the visual arts for students to create a visual illustrating the definitions of *amplitude, diffraction, refraction, pitch,* and *intensity*.

Grades 9–12

▷ Create an interest survey to evaluate student interest in the arts and apply the survey results to differentiate the use of the arts in classroom curriculum. For example, if a student has interest and experience in theater, ask that student to assist in creating a murder mystery using propositional logic, arguments, and methods of proof to find the culprit.

Jane Piirto (2014), editor of *Organic Creativity in the Classroom: Teaching Intuition in Academics and the Arts*, offered an innovative approach to integrating creativity and the arts in STEM classrooms. Instead of teaching the traditional approach of divergent thinking using fluency and flexibility, contributing authors in the edited volume focused on how to be creative in specific domains by using intuition. The following examples illustrated how to infuse creativity in the math and science classroom:

▷ Mathematics
- Encourage students to take risks by allowing them to discover multiple paths when finding the solution of math problems.
- Begin a unit by having a class discussion about a word that has a meaning both mathematically and in everyday life, which allows students to make connections between existing knowledge and the mathematical meaning.
- Provide time to solve a problem and complete an assignment. An incubation period of 3–5 days will allow students time to process material for a thoughtful understanding.
- Enhance intrinsic motivation by giving students brainteasers, puzzles, and games after completing a test.
- Provide opportunities for students to create personal definitions and interpretations about concepts before the definitions are taught in class.
- Encourage students to create their own problems to solve at the end of units.
- Provide opportunities for students to use their imagination and imagery. For example:
  - Have students create board games that incorporate and teach mathematical rules and concepts (e.g., a game that teaches about the background of postulates and theorems in geometry).
  - Give students play dough to create cross-sections of a solid and then calculate the volume of that solid.

▷ Science
- Provide opportunities for students to design and create innovative products (e.g., designing and building Rube Goldberg machines, automobiles, buildings and various structures, rockets, circuit boards, mobile phone applications).
- Allow students to create independent research projects in which students identify a problem they would like to solve and create a laboratory experiment to research and explore the problem.
- Allow students to create a play that teaches a concept (e.g., create a play that teaches the audience about fundamental particles and particle interactions).
- Use creative writing by asking students to write a science fiction story integrating specific scientific concepts learned in a unit.

The Kennedy Center for the Performing Arts offers a vast library of resources that showcases ways the arts can be infused into math and science for gifted and talented students. Table 13.1 includes examples of STEAM lessons that are offered by the Kennedy Center for the Performing Arts, all of which would be appropriate for gifted and talented students at the grade levels indicated. More resources are available online at the website displayed with alignment to language arts standards nationally and by state.

STEAM Grant Opportunity

Sometimes teachers of gifted and talented students may experience a lack of materials or resources to purchase additional materials needed to integrate the arts into current curriculum. Crayola (2013) and the National Association of Elementary School Principals fund a yearly grant program called Champion Creatively Alive Children that awards schools $3,500 to implement an innovative arts-integrated program. A maximum of 20 grants are awarded yearly. The application can be found on the following website: http://www.crayola.com/for-educators/ccac-landing/grant-program.aspx.

Suggested Recommendations for STEAM

Based on this review of the importance of innovation and the benefits of STEAM, the following suggestions can further assist educators of the gifted and talented in using STEAM effectively in their schools.

▷ Administrators of advanced programming can provide appropriate STEAM opportunities by setting up collaborations with teaching artists, inviting an arts peer coach, and exploring workshops and university courses focused upon integrating the arts.

▷ Teachers of the gifted and talented can seek out instructors of the arts (i.e., music, art, drama) and brainstorm how to further integrate the arts in classrooms.

▷ Teachers of the gifted and talented can collaborate with art instructors to create a concept-based thematic unit and team-teach relevant aspects of the unit.

TABLE 13.1
Sample STEAM Lessons from The Kennedy Center for the Performing Arts

Grade Levels	Content Areas	Synopsis of Lesson	Concepts Taught	Website
K–2	Math, Visual Arts	Students will create original comic strips to demonstrate their understanding of mathematical concepts.	History of comic strips, Literary elements of comic strips, any mathematical concepts	https://artsedge.kennedy-center.org/educators/lessons/grade-3-4/Creating_Comic_Strips
K–2	Science, Geography, Visual Arts	Students will paint original artwork to illustrate their understanding of different weather conditions.	Characteristics of different weather conditions, color, form, line, texture, space	https://artsedge.kennedy-center.org/educators/lessons/grade-3-4/Exploring_Weather#Overview
3–4	Science, Visual Arts	Students will create scaled mobiles of the solar system.	Force, distance, lever, mobile design of Alexander Calder	https://artsedge.kennedy-center.org/educators/lessons/grade-6-8/Planets_in_Motion#Overview
3–4	Math, Music	Students will compose original musical pieces that showcase their understanding of musical symbols and concepts.	Fractions, different types of musical notes and rests, bar lines, signature	https://artsedge.kennedy-center.org/educators/lessons/grade-6-8/First_Rhythmic_Composition
5–8	Science, Visual Arts, Media Arts	Students will create original firework displays to explore various chemical reactions.	Chemical reactions, repetition, emphasis balance	https://artsedge.kennedy-center.org/educators/lessons/grade-9-12/Oxidation_and_Combustion
5–8	Math, Music	Students will compose original musical pieces that use the Fibonacci sequence.	Fibonacci sequence, Phi, Golden Rule, music notation, scale, melody, harmony	https://artsedge.kennedy-center.org/educators/lessons/grade-9-12/Fibonacci_Music

▷ Teachers of the gifted and talented can provide opportunities for students to create and innovate inside the classroom, in schoolwide events, and in out-of-school venues.

▷ Teachers of the gifted and talented can seek out professional development opportunities to learn more about the arts and the application of the arts in the classroom. For example, *Champion Creatively Alive Children* (Crayola, 2013) is a free professional development program that teachers can complete at their own pace that offers strategies, resources, and tools to help educators advocate for arts integration.

▷ Researchers can create a valid and reliable instrument that would evaluate STEM teachers' interests in STEAM and their attitudes about integrating the arts into STEM curriculum. STEAM is a fairly new field of research, and there are currently no psychometrically sound STEAM instruments. This tool would be extremely beneficial in examining STEM teacher attitudes, and could be used as a foundational tool when providing professional development to STEM teachers interested in adding the arts to the curriculum.

Conclusion

Let's continue to gain STEAM with STEM! With the release of the Common Core State Standards (CCSS; National Governors Association Center for Best Practices & Council of Chief State School Officers, 2010), Next Generation Science Standards (NGSS; Achieve, Inc., 2013), and 21st Century Skills (Partnership for the 21st Century Skills, 2011), it has become increasingly important to integrate curriculum and standards across disciplines. To respond to the national STEM movement and the continual focus on literacy in the elementary grades, teachers who create interdisciplinary programming will encourage high-ability students to think both divergently and convergently (Maeda, 2013). Viewing curriculum and instruction through the lens of STEAM can serve as an effective method to approach the standards and develop technical thinkers with a range of abilities to bridge the technical and artistic domains for future innovation.

Discussion Questions

1. Reflect upon and discuss the benefits of integrating the arts into STEM described in the chapter.
2. List and describe three ways a classroom teacher at each school level could infuse the arts into a STEM lesson.
3. When a school offers a STEAM curriculum for the first time, what planning steps are important for administrators and teachers to consider and implement at each stage?

References

Achieve, Inc. (2013). *Next Generation Science Standards*. Washington, DC: Author.

Alvarez, L. (1987). *Adventures of a physicist*. New York, NY: Basic Books.

Andrews, B. W. (1999). Side by side: Evaluating a partnership program. *International Electronic Journal of Leadership in Learning, 3*(16) 1–22.

Andrews, B. W. (2006). Re-play: Re-assessing the effectiveness of arts education partnerships. *International Review of Education, 55*(2), 443–459.

Andrews, B. W. (2008). The Odyssey Project: Fostering teacher learning in the arts. *International Journal of Education and the Arts, 9*(11), 1–10.

Andrews, B. W. (2010). Seeking harmony: Teachers' Perspectives on learning to teach in and through the arts. *Encounters on Education, 11,* 81–98.

Barry, A. (2010). Engaging 21st century learners: A multidisciplinary, multiliteracy art-museum experience. *Studies on Learning, Evaluation, Innovation, and Development, 7*(3), 49–64.

Burnaford, G., & Scripp, L. (2013). Partnerships in arts integration research (PAIR) project (Final Reports). Retrieved from http://www.capeweb.org/formal-research-findings#

Catterall, J. S., Dumais, S. A., & Hampden-Thompson, G. (2012). *The arts and achievement in at-risk youth: Findings from four longitudinal studies*, Research Report #55. Washington, DC: National Endowment for the Arts.

Coxon, S. (2012). Innovative allies: Spatial and creative abilities. *Gifted Child Today, 35*(4), 277–284.

Crayola. (2013). *Arts-infused education leaderships: Videos and workshops* [Training video].

DeBoer, G., Carman, E., & Lazzaro, C. (2010). *The role of language arts in a successful STEM education program.* Retrieved from http://research.collegeboard.org/publications/content/2012/05/role-language-arts-successful-k-12-stem-implementation

Fantauzzacoffin, J., Rogers, J. D., & Bolter, J. D. (2012). From STEAM research to education: An integrated art and engineering course at Georgia Tech. *Proceedings of the IEEE Integrated STEM Education Conference* (ISEC 2012).

Furnham, A., Batey, M., Booth, T. W., Patel, V., & Lozinskaya, D. (2011). Individual difference predictors of creativity in art and science students. *Thinking Skills and Creativity, 6*(2), 114–121.

Garcia, L. (2003). The stories of pre-service theatre teachers who "resist". *Youth Theatre Journal, 17,* 1–16.

Higher Education Research Institute. (2010). *Degrees of success: Bachelor's degree completion rates among initial STEM majors.* HERI/CIRP Research Brief, January.

Johnson, K. (2014, Jan/Feb) Art- Science Approach for Gifted Learners. *Principal,* 42–43.

Kelly, S. N. (Oct. 2012). Fine arts-related instruction's influence on academic success. *Florida Music Director.* Retrieved from http://cfaefl.org/dnn/Portals/cfae/advocacy/2010-2011%20Cohort%20Study.pdf

Kerr, B., & McKay, R. (2013) Searching for tomorrow's innovators: Profiling creative adolescents. *Creativity Research Journal, 25*(1), 21–32.

Lamont, T. (2010, Nov. 13). John Maeda: Innovation is born when arts meets science. *The Observer.* Retrieved from http://www.theguardian.com/technology/2010/nov/14/my-bright-idea-john-maeda

Lynch, P. (2007). Making meaning many ways: An exploratory look at integrating the arts with classroom curriculum. *Art Education, 60*(4), 33–38.

Maeda, J. (2013). STEM + Art = STEAM. *The STEAM Journal, 1*(1), 1–2.

Moriwaki, K., Campbell, L., Brucker-Cohen, J., Saavedra, J., Stark, L., & Taylor, L. (2012). Scrapyard Challenge Jr., adapting an art and design workshop to support STEM to STEAM learning experiences. In *Proceedings of the IEEE Integrated STEM Education Conference* (ISEC 2012).

Moroye, C. M. (2009). *Complementary curriculum: The work of ecologically minded teachers.* New York, NY: Routledge.

National Governors Association Center for Best Practices, & Council of Chief State School Officers. (2010). *Common Core State Standards.* Washington, DC: Authors.

National Science Board. (2006). *Science and Engineering Indicators 2006.* Arlington, VA: National Science Foundation.

National Standards for Arts Education. (2014). Retrieved from https://artsedge.kennedy-center.org/educators/standards.aspx

Oreck, B. A. (2000). Teaching with arts survey. Unpublished survey. Storrs: University of Connecticut.

Oreck, B. A. (2004). The artistic and professional development of teachers: A study of teachers' attitudes toward and use of the arts in teaching. *Journal of Teacher Education, 55*(1), 55–69.

Oreck, B. A. (2006). Artistic choices: A study of teachers who use the arts in the classroom. *International Journal of Education & the Arts, 7*(8), 1–26.

Patteson, A. (2002). Amazing grace and powerful medicine: A case study of an elementary teacher and the arts. *Canadian Journal of Education, 27*(2/3), 269–289.

Partnership for 21st Century Skills. (2011). *Framework for 21st century learning.* Retrieved from http://www.p21.org/overview

Piirto, J. (2004). *Understanding creativity.* Scottsdale, AZ: Great Potential Press.

Piirto, J. (2011). *Creativity for 21st century skills: How to embed creativity into the curriculum.* Rotterdam, The Netherlands: Sense Publishers.

Piirto, J. (2014). *Organic creativity in the classroom: Teaching to intuition in academics and the arts.* Waco, TX: Prufrock Press.

Rinne, L., Gregory, E., Yarmolinskaya, J., & Hardiman, M. (2011). Why arts integration improves long-term retention of content. *Mind, Brain, and Education, 5*(2), 89–96.

Root-Bernstein, R., & Root-Bernstein, M. (2013). The art and craft. *Educational Leadership, 70*(5), 16–21.

Smithrim, K., & Upitis, R. (2005). Learning through the arts: Lessons of engagement. *Canadian Journal of Education, 289*(1/2), 109–127.

Sousa, D. A., & Pilecki, T. (2013). *From STEM to STEAM: Using brain-compatible strategies to integrate the arts.* Thousand Oaks, CA: Sage Publications.

Suzuki, S. (1969). *Nurtured by love: A new approach to education.* Hicksville, NY: Exposition.

Tomlinson, C. A., & Imbeau, M. B. (2010). *Leading and managing a differentiated classroom.* Alexandria, VA: ASCD.

Whisman, H., & Hixman, N. (2012). *A cohort study of arts participation and academic performance*. Charleston: West Virginia Department of Education, Division of Curriculum and Instructional Services, Office of Research.

CHAPTER 14

Neuroscience and Gender Issues in STEM Education and Among High-Ability Learners

Barbara A. Kerr, Ph.D., & J. D. Wright

The psychology of sex differences has a long history—almost as long as the study of psychology itself (Tavris, 1993). Psychologists have sought to understand the differences in behavior that people observe in females and males, and the public has long had a fascination with the origins of sex differences in abilities, personality, interests, and interpersonal behavior. Alice Eagly (1995) studied the dilemma that psychologists face when they enter into the arena of studying sex differences: "The political relevance of this work becomes apparent when it attracts media attention and is quickly incorporated into discourse on the status of women in society" (p.1). Often, the only access to research about sex differences that teachers have is from the popular media. Unfortunately, the popular media overemphasize small differences found, take leaps of the imagination in considering the implications, and often take any sex differences found as an endorsement of the status quo. In American society, females are expected to be less competent in STEM fields, less achieving in careers, and more oriented toward nurturing and relationships while males are expected to be more competent in STEM and career achievements, more aggressive, and less competent in caring and relationships (Fine, 2010). This chapter explores the history of sex differences studies and the newest frontier of sex differences

 DOI: 10.4324/9781003238218-16

research, neuroscience. It examines the flaws in research design and research-ers' interpretation of results, as well as the ways in which the media exaggerate and misinform the public. Finally, this chapter shows how the treatment of sex differences in psychometric and brain studies undermine the talent development of gifted boys and girls, through stereotype threat, gendered education, inappropriate career guidance, and lack of support for nontraditional paths to fulfillment of potential.

History of Sex Differences Research

A major theme of sex differences research has been that biological sex differences lead to inherent, immutable, and permanent differences between females and males. It is assumed that sex differences in abilities and personality are *inherent* in that simply being born a biological girl or a biological boy means that one also inherits, along with XX chromosomes or XY chromosomes, a host of other inherited characteristics that will determine what the child is capable of and how the child will behave. Sex differences research assumes that these inherited characteristics are *immutable* in that they cannot be changed by the environment—or only changed with great difficulty. Finally, this research often assumes that hormonally triggered changes that took place in the brain during the prenatal period are *permanent* in that one cascade of hormones is expected to induce changes that no further hormonal input can alter or redirect. All of these assumptions are wrong; yet they underlie not only most of sex differences research, but also the popular understanding of the brain and the development of abilities and personality traits (Fine, 2010).

Do sex chromosomes come laden with information about abilities and personality so that certain talents and characteristics are inherent in boys and girls? Of the tens of thousands of possible bits of information that genes carry, relatively few are sex-linked—and only rare cases of genetic anomalies have long-term consequences for development. Y-linked genes are very rare. Are they immutable? From the moment of conception, genes are in a constant dance with the intrauterine and extrauterine environment, gene expression is affected moment by moment by the input of the environment. The timing, location, and amount of gene expression affect the functions of the gene in a cell or in a multi-cellular organism. In addition, a gene's sex bias is not fixed, but can vary among tissues or change over the course of development (Parsch & Ellegren, 2013). Being genetically similar, twins have striking physical similarities, but what is most surprising to parents is their psychological dissimilarities—because even before the twin babies are born, the complex human brain has been changing constantly due to slight environmental changes (Kaplan & Rogers, 2003).

Over time, the presumed mechanisms by which genetic differences lead to behavioral differences change depending on the dominant theory of the period (Shields, 1975). Early physiologists such as Broca claimed it was brain size differences that led to males' greater accomplishments until it was pointed out that female brains were actually larger proportionate to their size. Then psychologists focused on relative differences in ratios, brain weight to height, to size of the heart, even to length of the femur. Every possible ratio was tried (Fine, 2013). When studies of mean differences in size or proportions of the brain devoted to "reason" didn't pan out (that is, female and male brains were found to be more alike than different), then sex hormones rose to prominence.

What about the "permanent" changes supposedly brought on by intra-uterine hormones? Because prenatal hormones at several stages of fetal development clearly are related to the development of female or male gonads, they seemed to sex differences researchers as the ideal candidates for explaining differences in toy preferences, interests, and later occupations of men and women. Geshwind proposed a hypothesis that Fine (2013) called the "Teflon pan" of hypotheses: No matter how much evidence refutes it, it doesn't stick. He proposed that fetal testosterone shrinks the left hemisphere of the male brain, leaving the right hemisphere dominant—and *therefore*—explaining why males are superior in right-hemisphere activities such as art, music, and mathematics. Besides the obvious flaws in the now debunked concept of a clear demarcation of right- and left-brain activities and in the difficulties of measuring accurately the amount and actions of fetal testosterone from mothers' blood and amniotic fluids, there is a more basic problem with the hypothesis. Both large postmortem studies of fetal brains and imaging studies of newborns show no evidence of a smaller left hemisphere in males (Fine, 2013). The hypothesis lives on, however, in the highly popularized work of Simon Baron-Cohen's *The Essential Difference* (2003). Baron-Cohen claimed that fetal testosterone creates essential brain differences leading to female brains that empathize and male brains that systemize, with both males and females along a spectrum from extreme female brain empathizers to extreme male brains, which he identifies as autistic brains. Baron-Cohen's evidence comes from his and his colleagues' studies showing that low testosterone brains of neonates look at faces more than high testosterone brains, and that low testosterone children show more preferences for "feminine toys" that suggest nurturing, and high testosterone children prefer "masculine toys" that encourage systematizing. Fine (2010) pointed out that the experiments were riddled with bias (in one, the experimenter herself was the face; in another, toys were named as masculine and feminine simply according to tradition rather than specific characteristics). In addition, she says, amniotic testosterone has not been found to be associated with any of the major "systemizing tests," whether these are spatial-rotation tests, tests of block design, or categorization. It is a clear case of another Teflon hypothesis, in that claims continue to be made of "hard-wired" brain differences as a result of hormones, for which evidence is conflicting, contrary, or absent.

One line of research in neuropsychology that has received significant press (Barford, 2014; Wolchover, 2012) explores childrens' toy preferences as a proxy for innate masculine or feminine interests. Alexander (2003) reviewed a number of studies that looked at the toy preferences of both human children and nonhuman primates, which found preferences for sex-typed toys for both males and females, and applied an evolutionary lens to the findings. He suggested that the tendency for male children, as well as female children exposed to increased levels of androgen in utero, to spend more time playing with "masculine" toys, such as cars or balls, could be the expression of an innate movement toward a more masculine gender role. Similarly, girls' preferences for "feminine" toys, such as dolls, could be the manifestation of intrinsic feminine interests and gender roles. These conclusions can be used to support the idea that boys and girls are genetically designed for different interests and career pathways, potentially allowing people to explain away gender gaps in career and achievement.

Another popular trend in gender differences research has been the use of finger length ratios (2D:4D) as an marker of prenatal testosterone exposure. Lutchmaya et al. (2003) looked at the ratio of testosterone to estradiol (the primary male and female sex hormones, respectively) present in the amniotic fluid of second trimester fetuses and compared that ratio to the 2D:4D finger length ratio of the same children at age 2. They found support for the belief that higher levels of prenatal testosterone were related to smaller (more masculine) 2D:4D ratios. With a cheap and simple measure to represent fetal testosterone levels in hand, researchers have been churning out new research finding what they claim are differences related to prenatal testosterone. In the past 10 years, 2D:4D has been correlated to all realms of differences, such as: aggression, risk aversion, spatial and navigational abilities, verbal fluency, and even SAT scores (e.g. Bailey & Hurd, 2005; Benderlioglu & Nelson, 2004; Kempel et al., 2005). Although on the surface many of the results seem to support traditional sex-differences research, the disconnect between what is actually being studied and the generalizations that are being drawn from the results is frequently ignored. In this case, finger length ratios are serving as a proxy for fetal testosterone levels, which in themselves are difficult to accurately measure, and which then serve as a proxy for sex or gender roles. The findings of these studies may be significant, but using those findings in media or popular press as support of traditional sex-role stereotypes, such as masculine individuals being more aggressive or having better spatial skills, constitutes a substantial leap from the actual data.

In an even more in-depth analysis of these studies, Jordan-Young (2010) questioned the very basis of brain organization theory that states that hormones organize the brain into permanent sex-typed structures. Re-examining and synthesizing all the studies on females with congenital adrenal hyperplasia (CAH), a condition in which female fetuses receive unusually high amounts of testosterone in utero, Young showed how these studies, which explored topics ranging from sexuality to play interests of these subjects, were flawed by a binary interpretation of hormones as masculine or feminine. Testosterone, estrogen, and other hormones have complex and often contradictory effects on physiology, brain functions, and behaviors depending on context—that is, the environment is constantly interacting with genes and hormones to affect all of these domains. Increased testosterone does not always lead to "masculine" structures and behaviors, and increased estrogen does not necessarily lead to "feminine" behaviors, and sometimes, in fact, they seem to be linked to the opposite behaviors. As a result, any studies that attempt to make a link between quasi-experiments on people with hormonal deviations versus "normal" people and some distal behavior like interests and preferences are based on the faulty equating of testosterone to masculine behavior and estrogen to feminine behavior. Unfortunately, however, extensive reviews of published articles show a strong bias toward citing those that show differences, and ignoring those that do not (Jordan-Young, 2010).

It is not only in brain studies, but in psychometric studies as well, that sex differences have been a constant theme. Sex differences have been studied in intelligence; math, verbal, spatial-visual, and various specific abilities; personality; interests; and a wide variety of interpersonal behaviors. From the beginning of psychometric studies in the early 1900s, at a time when sex roles for men and women were much more rigid, differences were indeed found in abilities and personality traits, and these differences were assumed to be related to immutable biological differences in males and females. By the 1970s, Maccoby and Jacklin (1974) had noticed that a great many studies that had found sex differences had not been published. They gathered all well-designed studies, published and unpublished on sex differences in intellect and social behaviors and carefully laid out the results in multiple tables. They found only a few consistent differences, such as higher math achievement and aggression in boys and higher verbal achievement and social interests in girls, but they found that for the most part, there was a great overlap in the distributions of traits and behaviors. They concluded that sex role socialization accounted for most of the intellectual and social differences found in boys and girls.

Early psychologists sought other means of supporting the idea of innate sex differences, beyond simply comparing means of females and males. Johann Meckel proposed the "variability hypothesis" to explain why more males were at both the lowest and highest ends of the distribution of scores on mental abilities—that is, why there were more males among the "mentally deficient" as well as more geniuses (Shields, 1975). Most eminent psychologists of the 20th century, including James McKeen Cattell and Edward Thorndike believed that the greater variability found in males was evidence of their superiority to females. The earliest researcher to take on the variability hypothesis was Leta Hollingworth (1914), who pointed out how errors in sampling led to fewer low-intelligence females being identified because they were less likely to be institutionalized, and fewer eminent females being identified because of social barriers to achievement. The variability hypothesis has cropped up repeatedly, however, as a way of explaining disproportionate numbers of males at the end of the curve in specific abilities and achievements, most recently as a way to explain the higher proportions of males to be found in the highest levels of math achievement.

When considering the findings from both the study of mean differences and variability differences between the sexes, however, changing social and cultural factors cast doubt on the findings of male superiority. Over the decades, as sex roles changed, so did the actual numbers of girls and women participating in high levels of educational attainment and leadership and in math and science achievement. The mathematics gap shrank and then disappeared, the higher education gap was reversed, and women began entering traditionally masculine fields in unprecedented numbers (Hyde, 2014).

The history of sex differences research in brain and psychometric studies foreshadows the continuation of many of the same themes in neuroscience. Colorful functional magnetic resonance imaging (fMRI) pictures of male and female brains, as opposed to black and white charts and diagrams, now illustrate a wide variety of scholarly and popular articles, and neuroscience has attained a higher status in academe and more glamour in the popular press than psychometric studies achieved, but the same research patterns and faults remain.

Neurotechnology and Neurosexism

Functional neuroimaging (FNI) includes fMRI, PET scans, and other methods that explore brain functions by imaging live brains while they are engaged in various mental tasks. Like all other technologies, it has been used too often to reinforce traditional gender roles and stereotypes in ways that are not justified by the actual data that has been produced (Fine, 2010). Among the popular "findings" of neuroscience that are accepted by the popular media as true are that males have more connections within hemispheres and females have more neural connections between hemispheres, that females have more activation of mirror neurons than males when confronted with pictures of people in distress, and that male brains have more capacity for spatial reasoning because of larger left hemispheres. Small samples give rise to unreliable, spurious results and poorly justified "reverse inferences" (i.e., claims of stereotype-consistent psychological differences between the sexes on the basis of brain differences). Fine (2010) also demonstrated how already weak neuroscientific conclusions are then grossly overblown by popular writers.

In exploring the sensationalizing effects of popular press on weak, many times flawed, research conclusions, Fine also discussed nonneuroimaging evidence cited as support for innate differences between the sexes. In one example, she detailed a number of prominent weaknesses in a study by Jennifer Connellan and colleagues (2000), which was widely cited. One and a half day-old babies were tested for preferences between a face and a mobile. Rather than presenting the stimuli simultaneously, as is standard practice when researching infant preferences, the stimuli were presented in sequence. The babies were also tested in different viewing positions, some horizontal on their backs and some held in a parent's lap, which could have affected their perceptions. Also, inadequate efforts were made to ensure the sex of the subject was unknown to the tester at the time of the test, which was even more crucial as Connellan, the lead author, acted as the face stimuli for testing. Even the interpretation of the data that was obtained could be seriously questioned. Finally, the authors assumed, without justification, that newborn looking preferences are a reliable "flag" for later social skills that are the product of a long and complex developmental process. Because of the degree to which the media seized upon these and similar findings, and the public embraced them, the results had far-reaching ripple effects on parents', educators', and researchers' understanding of sex differences, and the fact that the original studies were not well designed and the data not well interpreted serves to undermine another generation of boys and girls raised under false assumptions of behavioral and talent differences.

This history of sex differences is important because it illustrates recurring themes:

▷ an assumption that inherited sex differences in the brain account for adult differences in achievement, work, and relationships;

▷ an assumption that once biological differences are found, they trump development and environmental influences in determining adult behavior;

▷ the failure to find more than a few mean differences because of extreme overlap in the distribution of abilities and characteristics of males and females;

▷ the revelation of flawed sampling, design, and interpretation of results in many studies; and

▷ the finding of a few differences leading to rekindled interest, new methods, and new claims for the existence of inherent differences (Fine, 2010; Kerr & McKay, 2014).

None of these arguments are meant to deny that there are any sex differences beyond sex organs and secondary sexual characteristics; in medical research, it has become clear that it is important to understand where biologically based differences lead to the need for different treatments for men and women. Educators, however, need to be wary of any research that purports to make the case that innate brain differences require different kinds of educational treatments, because there is simply not enough neuroscience evidence to warrant such an "essentialist" assumption (Cohen, 2009). Instead, educators need to consider how powerful sociocultural influences contribute to stereotypes that limit both female and male potential, and how schooling may contribute to the development of stereotypes.

One of the largest scale studies to explore the relationship that cultural stereotypes have on gender academic achievement gaps in science and math was conducted by Nosek et al. in 2009. The authors used data from close to 300,000 respondents across 34 countries on a measure of implicit and explicit gender science stereotypes. The implicit measure looked at the ease with which respondents paired science-related words (e.g., *science, chemistry, physics*) or liberal arts-related words (e.g., *arts, history*) with male or female words (e.g., *he, him, she, her*). More than 70% of respondents worldwide are significantly faster at matching male words with science words and female words with liberal arts words, as opposed to men with liberal arts and women with science. This is interpreted as an overall implicit stereotype linking men with science and women with language arts.

> More than 70% of respondents worldwide are significantly faster at matching male words with science words and female words with liberal arts words, as opposed to men with liberal arts and women with science. This is interpreted as an overall implicit stereotype linking men with science and women with language arts (Nosek et al., 2009).

Nosek et al. (2009) then compared gender science implicit stereotypes across various countries with gender differences in science achievement for eighth graders in each country. They found that, as expected, increased implicit stereotypes were associated with larger gender gaps, but impressively, that association was still found to be significant when controlling for explicit stereotypes, GDP, and a gender gap index (GPI). Even though the IAT looked only at science stereotypes, the authors also found significant correlations related to gender differences in math achievement. The fact that gender differences in science and math achievement can be explained to such a degree by cultural factors implies that changes in beliefs can impact achievement.

Sex Differences and Gifted Girls

Whereas public opinion continues to hold that innate sex differences in cognitive abilities account for differential achievement of males and females, repeated meta-analyses have found no compelling evidence of this (Halpern, 2013; Hyde, 2014). Interestingly, however, sex differences continue to hold a fascination for scholars in gifted education. As soon as it was clear that the gap between math achievement scores for boys and girls had closed, researchers quickly turned attention to other abilities that might show differences between sexes such as spatial-visual abilities (Halpern, 2013; Tavris, 1993). The question should be not be *what are the important sex differences*, but *why are sex differences so interesting to gifted education researchers*? What impact does such fascination with sex differences have on gifted girls' and gifted boys' ideas of what kind of life is possible for them?

Whereas public opinion continues to hold that innate sex differences in cognitive abilities account for differential achievement of males and females, repeated meta-analyses have found no compelling evidence of this (Halpern, 2013; Hyde 2014).

Quantitative studies tend to be investigations of sex differences between gifted boys and girls, usually comparisons of abilities (Dai, 2002; Swiatek, Lupkowski, & O'Donoghue, 2000). Some authors ask if the topic of gifted girls' underachievement is obsolete (Schober, Rieman, & Wagner, 2004). Many seem to assume that any special issues that gifted females once had—lower math and science achievement, lower self-esteem, and lower career accomplishments compared to gifted males—have been addressed and resolved. An extensive examination of the literature by Dai (2002) pointed out that there are still many questions unanswered about the development of gifted girls.

Indeed, great progress has been made in narrowing the math and science gap. Not only has the gender gap in math achievement scores closed among girls and boys in general (Hyde, 2005; Hyde et al;, 2008), the once very wide gap at the highest levels of ability has narrowed significantly (Brody & Mills, 2005; Halpern, 2013). Women are also entering many fields, including social sciences, health sciences, natural sciences, business, and law, in equal or greater proportions as men (Kerr & McKay, 2014). Young women, however, enter and persist in the physical sciences, engineering, and computer science at much lower rates than males (National Science Foundation, 2009). Studies of valedictorians also show a still disturbing tendency for females to go to less prestigious colleges, to avoid physical and computer sciences, and to enter lower paying careers (York, 2008). Therefore, the changes in girls' achievement scores have not always translated into higher aspirations and career accomplishments.

Studies of valedictorians also show a still disturbing tendency for females to go to less prestigious colleges, to avoid physical and computer sciences, and to enter lower paying careers (York, 2008). Therefore, the changes in girls' achievement scores has not always translated into higher aspirations and career accomplishments.

Misperceptions of the level of ability, or "brilliance" needed for various disciplines may underlie gender distributions in academic fields (Leslie, Cimpian, Meyer, & Freeland, 2015). When media images and career information material focus on the highest levels of abilities, and when the face of genius is male, gifted girls may come to believe that they lack the abilities needed to succeed in the field in which they are interested.

Socialization of Gifted Girls

In *Smart Girls in the Twenty-First Century*, Kerr and McKay (2014) document the ways in which intelligence, personality, and privilege interact in the socialization of gifted girls at each stage of development. As infants, girls are helped and handled more; and as preschoolers, a frenzy of sex role socialization occurs with the onslaught of the "Princess Industrial Complex" (Orenstein, 2011). Young gifted girls often eagerly embrace the toys, clothes, and stories associated with the Disney princesses, learning that being pretty, popular, and romantically attractive are achievements worth striving for. Those bright girls who prefer books to dolls and Legos to kitchen sets may find themselves rejected by other children who are in the developmental stage in which their gender categories are strictly observed. In elementary school, bright girls may be discouraged from precocious reading, denied educational opportunities that are perceived as too challenging by protective parents, and exposed to sex role stereotypes in their textbooks, educational media, and play activities (Collins, 2011). Even though gifted girls often prefer to play with gifted boys or older girls, these options are often closed by segregated playgrounds and rigid age-in-grade progression (Kerr & McKay, 2014). By middle school, a culture of romance and premature sexualization of teen girls may create a false choice for gifted girls—to invest in romantic relationships or to invest in achievement of academic and career goals (Holland & Eisenhart, 1990). The impact of social media may accelerate teen girls' self-objectification (Doornwaard et al, 2014). In secondary school, although gifted girls continue to have high academic achievement in all domains, their self-efficacy in nontraditional areas may decline. The "confidence gap" exists despite comparable prior accomplishments, such as STEM course grades (Pajares, 2005), and is partly responsible for the "gender gap" in engineering and other STEM disciplines—including computer science, physics, and astronomy—that becomes apparent at colleges throughout the U.S.

> Even though gifted girls often prefer to play with gifted boys or older girls, these options are often closed by segregated playgrounds and rigid age-in-grade progression (Kerr & McKay, 2014) . . . . In secondary school, although gifted girls continue to have high academic achievement in all domains, their self-efficacy in nontraditional areas may decline. The "confidence gap" exists despite comparable prior accomplishments, such as STEM course grades (Pajares, 2005), and is partly responsible for the "gender gap" in engineering and other STEM disciplines—including computer science, physics, and astronomy—that becomes apparent at colleges throughout the U.S."

Media Reports and Beliefs About Ability

Throughout education, math-science self-efficacy affects girls' aspirations in science, technology, engineering, and mathematics (STEM) fields. For gifted girls, this is also the case (Dai, 2002). Many reasons have been proposed for girls' lower self-efficacy in math and science, with socialization, media influence, and stereotype threat (Spencer, Steele, & Quinn, 1999) considered the major sociocultural influences. Jacobs (Jacobs, 2005; Jacobs & Eccles, 1985, 1992) has followed this phenomenon for 25 years. An example of the power of the media to shape parental beliefs is the extensive media coverage of a study published in *Science* (Benbow & Stanley, 1980). They found sex differences in mathematics achievement favoring boys at the highest end of mathematics achievement on the SATs by students who took the college entrance exam when they were only in seventh grade as a part of a talent search program. Ignoring all nuance associated with a highly selective sample of U.S. students, headlines in major newsmagazines suggested that boys have a "math gene" missing by girls and that girls were just "naturally" less able than boys in math. Jacobs and Eccles (1985) had access to data on parental beliefs about their children's math abilities prior to the media blast about sex differences and after the event. On exactly the same measures, they found that mothers who had been exposed to media reports believed that their daughters had less ability in math, were less likely to succeed in math in the future, were more likely to find math difficult, and were more likely to have to work harder to succeed in math. Sadly, even 12 years after the study, upon follow-up, it was found that these mothers' beliefs still mattered. Mothers' earlier perceptions of their adolescents' abilities were related to adolescents' math-science self-efficacy 2 years after high school, with adolescents' self-perceptions of math ability during 10th grade mediating the relation with mothers' perceptions. Moreover, mothers' earlier predictions of their children's abilities to succeed in math careers were significantly related to later career choices, with girls who had mothers who had lower beliefs in their math ability less likely to enter math-intensive and physical sciences careers (Bleeker & Jacobs, 2004).

> Moreover, mothers' earlier predictions of their children's abilities to succeed in math careers were significantly related to later career choices, with girls who had mothers who had lower beliefs in their math ability less likely to enter math-intensive and physical sciences careers (Bleeker & Jacobs, 2004).

Stereotype Threat

Stereotype threat has been found to be a powerful influence on female and minority students' achievement on standardized tests. Stereotype threat occurs in situations where one can be judged by, treated in terms of, or self-fulfill negative stereotypes about one's group (Spencer, Steele, & Quinn, 1999). Spencer et al. (1999) stressed that it is not "peculiar to the internal psychology of particular groups and it can be experienced by the members of any group about whom negative stereotypes exist—generation 'X,' the elderly, white males, etc." (p. 6). This threat, when activated, has been shown to lead to changes in academic performance and behaviors (Deemer, Smith, Carroll, & Carpenter, 2014; Steele & Aronson, 1995), job behaviors (Roberson, Deitch, Brief, & Block, 2003), career decisions, and self-esteem (Deemer, Thoman, Chase, & Smith, 2014). Activation of the threat occurs most prominently when dealing with performance diagnostic situations and explicit statements of the relevant stereotype, but effects can also be seen to varying degrees with more subtle cues and in nondiagnostic situations (Steele & Aronson, 1995). The primary mechanism through which stereotype threats is proposed to affect performance is the reduction of working memory capacity, as the affected individual struggles to both complete a task and work to disconfirm a salient stereotype (Schmader & Johns, 2003).

A meta-analysis of stereotype threat effects (Nguyen & Ryan, 2008) found differential effects of race- versus gender-based stereotypes. Women experienced smaller performance decrements than did minorities when tests were difficult. For women, subtle threat-activating cues, such as simply having to identify one's gender prior to testing, produced the largest effect. Explicit threat-removal strategies, such as teaching students about the effects of stereotype threat, were more effective in reducing stereotype threat effects than subtle ones. For minorities, moderately explicit stereotype threat-activating cues, such as mentioning that Blacks typically perform worse on this type of task, produced the largest effect and explicit removal strategies actually enhanced stereotype threat effects compared with subtle strategies. In addition, stereotype threat affected moderately math-identified women more severely than highly math-identified women. These results could suggest that although women who are already strongly committed to a math or science field may be less susceptible to the deleterious effects of stereotype threat, women who have an interest in math or science, but are more undecided on their future academic path, are more vulnerable to those negative effects. This could help explain the erosion of women in STEM fields as one moves higher through the academic track.

To summarize, the math achievement gaps have for all intents and purposes closed, for girls in general and for gifted girls. Bright girls are entering traditionally male fields in large numbers, equaling proportions of males everywhere but in STEM fields. Nevertheless, self-efficacy continues to be a problem for gifted girls, with internalization of societal and parental attitudes affecting their progress in achieving their career goals. Sex differences research and the media reports that trumpet the results of neuroscience findings of sex differences may play an important role in activating stereotype threat and discouraging gifted girls from fulfilling their potential in STEM and other male stereotyped domains.

Sex Differences Research and Gifted Boys

In popular opinion, as well as neuropsychological research, aggression and emotional expression are believed to be domains in which some of the most pronounced differences between the genders exist. In her 2014 review of the literature, Hyde noted that consistent gender differences in aggression do seem to exist; however, the relationship is far more dependent on context than most people recognize. Males, for example, may be more likely to exhibit aggression when cued by violence but not directly provoked, but when both cued and provoked, women are just as likely as men to show aggression (Bettencourt & Kernahan, 1997). There is also research supporting the popular belief that boys tend to suppress overall emotional expression compared to girls, and that they are less likely to show positive emotions (Chaplin & Aldao, 2013; Polce-lyn & Myers, 1998). When taken together, teachers and parents may see a distressed or struggling boy in class, who only expresses himself through acting out or aggressive acts, as "just being a boy," rather than looking deeper for what needs may be driving his behavior. This presents an especially salient issue for gifted boys, who are more likely to be bored and frustrated in school, potentially channeling that energy into external behaviors, such as inattention and acting out (Kerr & Cohn, 2001). If social stereotypes keep teachers or parents from delving further into the issues, the boy may receive punishment, or even a diagnosis, rather than the support and challenge he needs.

Socialization of Gifted Boys

Boys are socialized from a very young age in "appropriate" sex roles, with kindergarten-aged children already having significant knowledge about sex stereotypes and male stereotypes being learned at an even younger age than female stereotypes (Williams, Bennett, & Best, 1975). By kindergarten, more than 90% of children studied were able to identify "aggressive" and "strong" as related to males rather than females. A major component of boys' socialization around masculinity is related to the avoidance of anything "feminine" or "girly" (Kerr & Cohn, 2001). For many gifted boys, the classmates they would be most likely to socialize with or feel a connection to are gifted girls, who are likely to be developmentally closer to them than other boys their age. Social pressures to avoid girls or girly behaviors, however, may keep many gifted boys from pursuing or accepting those interactions, instead finding themselves stuck in all-male peer groups where they feel isolated and are forced to put up a "manly" front.

Another challenge that gifted boys may face when trying to develop peer relationships is asynchronous development, frequently occurring when gifted boys' social and emotional development lags behind their cognitive or academic development (Silverman, 1997). Although the gifted boy may find intellectual peers in higher grades, it is likely that his emotional development is far enough below older boys' and girls' that friendships would be difficult at best. Thus, for many gifted boys, especially those in smaller communities where they may be the only gifted boys in their grade, they find themselves isolated from peer groups, either because of their more advanced intelligence or less advanced social skills. A potential peer group already sliced in half by sex-role pressures becomes dangerously thin for gifted boys when factoring in developmental differences as well.

Another of the socializations around masculinity that may have the potential to damage gifted boys' abilities to succeed and prosper in academic settings is that men should be self-sufficient to the point that asking for help is a sign of weakness. Males' reluctance to seek help for physical and mental health conditions has been well documented (Davies et al., 2000; Levant, Wimer, & Williams 2011), but research has also explored men and boys' unwillingness to seek help in academic settings (Kessels & Steinmayr, 2013). This is especially troublesome for gifted boys, who are less likely to receive adequate instruction without special support. Their exceptional abilities not only put them at risk for underperformance due to boredom, it also may increase the chances of bullying and social isolation (Kerr & Cohn, 2001). When combined with the masculine stereotypes around help-seeking behavior, gifted boys face a difficult catch-22; they can either suffer through the difficulties, missing potentially much needed support, or they can step even further out of the traditional "boys club" and ask for the help and assistance they may need.

Another potential consequence for gifted boys who either avoid seeking advice or cannot find the right advice is ending up on an academic or career path they simply do not like. Many gifted children experience multipotentiality in their academic and career options; they find themselves very good in a large number of subjects. To most of their teachers, counselors, and parents, this is hardly a negative, but the surplus of available options can be overwhelming to the gifted child (Kerr & Cohn, 2001). A lack of understanding and guidance around choosing a career based on values and interests, combined with stereotypes and enculturation saying that men are biologically "designed" to work in the STEM fields, may lead some gifted boys and adolescents to move toward career paths with which they may have no real connection. A gifted boy's passion and skill for interior design, teaching, art, or child care may be lost because he has been socialized to avoid traditionally feminine careers. Although he may have the grades and abilities for a more masculine career, if that pathway doesn't align with his values and interests, he may become dissatisfied and unlikely to excel or reach his highest potential.

Sex differences research and the broad (mis)interpretation of results may also serve to limit the talent development of gifted boys. Because differences have been found in help-seeking behavior, gifted boys may be less likely to use academic assistance, guidance, or therapy when they need it. Research that seems to show that boys have an advantage with abilities related to STEM may make it more difficult for boys who are very talented in math, science, and technology to make other career choices, even when they lack interest in those more traditionally masculine fields.

Suggestions for Parents, Educators, and Scholars

Some suggested guidelines for parents, educators, and scholars are as follows:

1. In discussions of gender, focus on similarities of boys and girls rather than differences, emphasizing that although there may be differences

in how boys and girls act and are portrayed in the media, gifted girls and gifted boys are more similar in abilities, personality, and career potential than they are different.

2.   Build awareness of sex role stereotypes and how they are promoted in conversations, in the media, and at school. Question claims of psychological differences as "hard-wired," unchangeable, or important to childrens' potential.

3.   Research that finds differences at the highest levels of abilities have great salience for gifted boys and girls, who may not understand these kinds of differences do not imply differences in career potentials. Career education should focus on the average level of ability required for particular high-level occupations, rather than implying that the highest levels of abilities in verbal, mathematical, or spatial-visual ability are needed for success.

4.   Build awareness of media exaggerations or misinterpretations of psychometric or neuroscience studies of sex differences by reading critiques of these studies and seeking out the original research to examine the actual findings.

5.   Train bright young people to know how to dispute sex role stereotypes, and prepare them for tests and assessments with positive statements about their capacity to succeed and demonstrate their knowledge, abilities, and positive characteristics.

6.   Teachers should carefully evaluate claims that boys and girls have different "learning styles" or specific educational needs based on innate differences, since scant evidence exists for these claims. Although there may be good reasons to provide different kinds of educational experiences, such as some single-sex options, these should be based on socio-cultural arguments and should be focused on expanding options rather than limiting options.

7.   Scholars should consider how their findings related to sex differences will be treated by the media, and make every effort to be sure that interpretations of their work are accurate and limited to the actual data.

8.   Findings based on neuroscience imaging should not be considered any more valid for drawing broad conclusions for education than findings based on any other technology. Although colorful illustrations of brains in action are fascinating, the imaging technology is not yet at a point where clear causal statements can be made about brain differences and behavioral differences of boys and girls and men and

women. Because gifted girls and boys are more likely to be avid readers of popular science, they need guidance in critical thinking about neuroscience as well as encouragement to dig deeper into the literature in order to build their understanding of how science is constructed and interpreted by society.

References

Alexander, G. M. (2003). An evolutionary perspective of sex-typed toy preferences: Pink, blue, and the brain. *Archives of Sexual Behavior, 32*(1), 7–14.

Bailey, A., & Hurd, P. (2005). Finger length ratio (2D:4D) correlates with physical aggression in men but not in women. *Biological Psychology, 68*(3), 215–222.

Barford, V. (2014, Jan. 26). Do children's toys influence their career choices? *BBC News Magazine.* Retrieved from http://www.bbc.com/news/magazine-25857895

Baron-Cohen, S. (2003). *The essential difference: Male and female brains and the truth about autism.* New York, NY: Basic Books.

Benbow, C. P., & Stanley, J. C. (1980). Sex differences in mathematical ability: Fact or artifact? *Science, 210*(4475), 1262–1264.

Benderlioglu, Z., & Nelson, R. (2004). Digit length ratios predict reactive aggression in women, but not in men. *Hormones and Behavior, 46*(5), 558–564.

Bettencourt, B. A., & Kernahan, C. (1997). A meta-analysis of aggression in the presence of violent cues: Effects of gender differences and aversive provocation. *Aggressive Behavior, 23*(6), 447–456.

Bleeker, M. M., & Jacobs, J. E. (2004). Achievement in math and science: Do mothers' beliefs matter 12 years later? *Journal of Educational Psychology, 96*(1), 97.

Brody, L. E., & Mills, C. J. (2005). Talent search research: what have we learned? *High Ability Studies, 16*(1), 97–111.

Chaplin, T., & Aldao, A. (2013). Gender differences in emotion expression in children: A meta-analytic review. *Psychological Bulletin, 139*(4), 735–765.

Cohen, D. S. (2009). No boy left behind? Single-sex education and the essentialist myth of masculinity. *Indiana Law Journal, 84,* 135.

Collins, R. L. (2011). Content analysis of gender roles in media: Where are we now and where should we go? *Sex Roles, 64*(3-4), 290–298.

Connellan, J., Baron-Cohen, S., Wheelwright, S., Batki, A., & Ahluwalia, J. (2000). Sex differences in human neonatal social perception. *Infant Behavior and Development,* 23(1), 113–118.

Dai, D. Y. (2002). Are gifted girls motivationally disadvantaged? Review, reflection, and redirection. *Journal for the Education of the Gifted,* 25(4), 315–358.

Davies, J., McCrae, B.P., Frank, J., Dochnahl, A., Pickering, T., Harrison, B., Zakrzewski, M., & Wilson, K. (2000). Identifying male college students' perceived health needs, barriers to seeking help, and recommendations to help men adopt healthier lifestyles. *Journal of American College Health,* 48(6), 259–267.

Deemer, E. D., Smith, J. L., Carroll, A. N., & Carpenter, J. P. (2014). Academic procrastination in STEM: Interactive effects of stereotype threat and achievement goals. *The Career Development Quarterly,* 62(2), 143–155.

Deemer, E. D., Thoman, D. B., Chase, J. P., & Smith, J. L. (2014). Feeling the threat: Stereotype threat as a contextual barrier to women's science career choice intentions. *Journal of Career Development,* 41(2), 141–158.

Doornwaard, S. M., Moreno, M. A., van den Eijnden, R. J., Vanwesenbeeck, I., & Ter Bogt, T. F. (2014). Young adolescents' sexual and romantic reference displays on Facebook. *Journal of Adolescent Health,* 55(4), 535–541.

Eagly, A. H. (1995). The science and politics of comparing women and men. *American Psychologist,* 50(3), 145.

Fine, C. (2010). *Delusions of gender: The real science behind sex differences.* London, England: Icon Books.

Fine, C. (2013). Is there neurosexism in functional neuroimaging investigations of sex differences? *Neuroethics, 6* (2), 369–409.

Halpern, D. F. (2013). *Sex differences in cognitive abilities.* London, England: Taylor & Francis.

Holland, D., & Eisenhart, M. A. (1990). Educated in romance. *Women, Achievement, and College.* Chicago, IL: University of Chicago Press.

Hollingworth, L. S. (1914). Variability as related to sex differences in achievement: A critique. *The American Journal of Sociology,* 19(4), 510–530.

Hyde, J. S. (2005). The gender similarities hypothesis. *American Psychologist,* 60(6), 581–592.

Hyde, J. S. (2014). Gender similarities and differences. *Annual Review of Psychology, 65,* 373–398.

Hyde, J. S., Lindberg, S. M., Linn, M. C., Ellis, A. B., & Williams, C. C. (2008). Gender similarities characterize math performance. *Science, 321*(5888), 494–495.

Jacobs, J. E. (2005). Twenty-five years of research on gender and ethnic differences in math and science career choices: What have we learned? *New Directions for Child and Adolescent Development, 2005*(110), 85–94.

Jacobs, J. E., & Eccles, J. S. (1985). Gender differences in math ability: The impact of media reports on parents. *Educational Researcher, 14*(3), 20–25.

Jordan-Young, R. (2010). *Brain storm: The flaws in the science of sex differences.* Cambridge, MA: Harvard University Press.

Kaplan, G., & Rogers, L. (2003). *Gene worship: Moving beyond the nature/nurture debate over genes, brain, and gender.* New York, NY: Other Press.

Kempel, P., Gohlke, B., Klempau, J., Zinsberger, P., Reuter, M., & Hennig, J. (2005). Second-to-fourth digit length, testosterone and spatial ability. *Intelligence, 33,* 215–230.

Kerr, B. A., & Cohn, S. J. (2001). *Smart boys: Talent, manhood, and the search for meaning.* Scottsdale, AZ: Great Potential Press.

Kerr, B. A., & McKay, R. A. (2014). *Smart girls in the twenty-first century.* Scottsdale, AZ: Great Potential Press.

Kessels, U., & Steinmayr, R. (2013). Macho-man in school: Toward the role of gender role self-concepts and help seeking in school performance. *Learning and Individual Differences, 23,* 234–240.

Leslie, S. J., Cimpian, A., Meyer, M., & Freeland, E. (2015). Expectations of brilliance underlie gender distributions across academic disciplines. *Science, 347*(6219), 262–265.

Levant, R. F., Wimer, D. J., & Williams, C. M. (2011). An evaluation of the Health Behavior Inventory-20 (HBI-20) and its relationships to masculinity and attitudes towards seeking psychological help among college men. *Psychology of Men & Masculinity, 12*(1), 26–41.

Lutchmaya, S., Baron-Cohen, S., Raggatt, P., Knickmeyer, R., & Manning, J. T. (2003). 2nd to 4th digit ratios, fetal testosterone and estradiol. *Early Human Development, 77,* 23–28.

Maccoby, E., & Jacklin, C. (1974). *The psychology of sex differences.* Stanford, CA: Stanford University Press.

National Science Foundation. (2009). Women, minorities, and persons with disabilities in science and engineering. Reston, VA: Author.

Nguyen, H. H. D., & Ryan, A. M. (2008). Does stereotype threat affect test performance of minorities and women? A meta-analysis of experimental evidence. *Journal of Applied Psychology, 93*(6), 13–14.

Nosek, B. A., Smyth, F. L., Sriram, N., Lindner, N. M., Devos, T., Ayala, A., & Greenwald, A. G. (2009). National differences in gender-science stereotypes predict national sex differences in science and math achievement. *Proceedings of the National Academy of Sciences of the United States of America, 106*(26), 10593–10597.

Orenstein, P. (2011). *Cinderella ate my daughter*. New York, NY: HarperCollins.

Pajares, F. (2005). Gender differences in mathematics self-efficacy beliefs. In A. M. Gallagher & J. C. Kaufman (Eds.), *Gender differences in mathematics: An integrative psychological approach* (pp. 294–315). New York, NY: Cambridge University Press.

Parsch, J., & Ellegren, H. (2013). The evolutionary causes and consequences of sex-biased gene expression. *Nature Reviews Genetics, 14*(2), 83–87.

Polce-lyn, M., & Myers, B. (1998). Gender and age patterns in emotional expression, body image, and self-esteem: A qualitative analysis. *Sex Roles, 38*(11/12), 1025–1048.

Roberson, L., Deitch, E. A., Brief, A. P., & Block, C. J. (2003). Stereotype threat and feedback seeking in the workplace. *Journal of Vocational Behavior, 62*(1), 176–188.

Schmader, T., & Johns, M. (2003). Converging evidence that stereotype threat reduces working memory capacity. *Journal of Personality and Social Psychology, 85*(3), 440–452.

Schober, B., Rieman, R., & Wagner, P. (2004). Is research on gender-specific underachievement in gifted girls an obsolete topic? New findings on an often discussed issue. *High Ability Studies, 15*(1), 43–62.

Shields, S. (1975). Functionalism, Darwinism, and the psychology of women. *American Psychologist, 30*(7), 739–754.

Silverman, L. K. (1997). The construct of asynchronous developement. *Peabody Journal of Education, 72*(3–4), 36–58.

Spencer, S. J., Steele, C. M., & Quinn, D. M. (1999). Stereotype threat and women's mathematics performance. *Journal of Experimental Social Psychology, 35*(1), 4–28.

Steele, C. M., & Aronson, J. (1995). Stereotype threat and the intellectual test performance of African Americans. *Journal of Personality and Social Psychology, 69*(5), 797–811.

Swiatek, M. A., Lupkowski-Shoplik, A., & O'Donoghue, C. (2000). Gender differences in above-level EXPLORE scores of gifted third through sixth graders. *Journal of Educational Psychology, 92*(4), 718.

Tavris, C. (1993). The mismeasure of woman. *Feminism & Psychology, 3*(2), 149–168.

Williams, J., Bennett, S., & Best, D. (1975). Awareness and expression of sex stereotypes in young children. *Developmental Psychology, 11*(5), 635–642.

Wolchover, N. (2012, Aug. 24). Gender & toys: Monkey study suggests hormonal basis for children's toy preferences. *Huffington Post*. Retrieved from http://www.huffingtonpost.com/2012/08/24/gender-toys-children-toy-preferences-hormones_n_1827727.html

York, E. A. (2008). Gender differences in the college and career aspirations of high school valedictorians. *Journal of Advanced Academics, 19*(4), 578–600.

CHAPTER 15

Designing and Developing an Exceptional Living-Learning Environment for the Highly Able Student

A Grassroots Approach to a Statewide STEM Initiative

Judy K. Stewart, Ph.D., Christopher R. Gareis, Ed.D., & M. Caroline Martin, RN, MHA

Introduction

Virginia is in an enviable position. Numerous studies suggest the Commonwealth is poised to offer upward of 400,000 STEM-related jobs as early as 2018 (Carnevale, Smith, & Strohl, 2010; Virginia Chamber of Commerce, 2013). And yet, there is a deficit of knowledge workers in the pipeline; a wide disparity in offerings for highly able students across its K–12 school divisions; and limited student exposure to rich, integrated science, technology, engineering, and applied mathematics curriculum early and continuously throughout students' academic careers (Change the Equation, 2014; Lederman, 1998).

Virginia's public school system is comprised of 133 school divisions, which are contiguous with the state's counties and cities. As is the case in many states, Virginia's public school divisions include some with very large student enrollments in the tens of thousands, as well as others with very small enrollments of fewer than one thousand. There are urban, suburban, and rural locales; well-resourced and poorly resourced divisions; and both high-performing and highly challenged school divisions. All too frequently, where a student lives determines

 DOI: 10.4324/9781003238218-17

a student's access and opportunity. If Virginia is to meet the need for 21st-century knowledge workers and sustain a position as an economic leader, it must invest in student talent. More to the point, it must invest in developing exceptionally able student talent, regardless of student zip code (Finn & Hockett, 2012).

To meet this challenge, a small, grassroots group of expert stakeholders met in November 2010 and formed the Virginia Science Technology Engineering and Applied Mathematics Academy, also known as the Virginia STEAM Academy. As individuals, members of the founding steering committee brought to the table expertise in education, operations, and nonprofit board leadership. Collectively, the steering committee represented varied experiences and perspectives from fields such as K–12 and higher education; local, state, and international business; science and engineering; and government.

Over the next 4 years, the Virginia STEAM Academy evolved into a statewide, nonprofit, charitable organization with a clearly articulated mission and vision. Encouraged by the example of more than a dozen publicly accessible, statewide, boarding, and STEM-focused high schools already in existence across the country, the group undertook efforts that will result in establishing a strong, statewide-representative board; acquiring resources; locating a campus home; developing a framework for curriculum, instruction, and assessment; and defining the profile and selection criteria for Virginia STEAM Academy faculty, staff, and students.

In this chapter, we share the seminal experiences and impactful accomplishments that have informed the design and development of the Virginia STEAM Academy to date. We also outline the key steps to bringing the full scope of the initiative to fruition. At the outset, we should recognize that our tone in this chapter is sometimes more rhetorical than academic. This is the case for two reasons: First, the story of the Virginia STEAM Academy is still being written; the work of developing, designing, and bringing to full manifestation this school is still underway. Therefore, in the vein of Kouzes and Posner (2002), such core leadership practices as *inspiring a shared vision, enabling others to act*, and *encouraging the heart* are still very much at play. The second reason is that two of the lead authors of this chapter are cofounders of the Virginia STEAM Academy and are therefore deeply committed to and invested in the school. We appreciate the opportunity to tell part of the story of the Virginia STEAM Academy at this point in its evolution. We hope that our experiences, lessons learned thus far, and plans for next steps in creating a publicly accessible, residential STEM school for highly able students will be instructive for others who aspire to pursue a similar vision.

Mission

The mission of the Virginia STEAM Academy is to nurture future generations of creative, ethical, and imaginative STEAM leaders who understand and integrate the humanities into their full development. Its purpose is to accelerate learning for highly able students, attract more students to STEAM disciplines early in their academic careers, and positively impact the Commonwealth of Virginia's economic future. To pursue this mission, the Academy is comprised of four components:

1. A publicly accessible, statewide, residential high school designed to serve approximately 500 students in grades 9–12;

2. A cost-free, statewide, residential summer academy intended to inspire and expose middle school students to STEAM content, applied learning, and leadership development;

3. High-quality professional development for teachers and administrators, not only at the Virginia STEAM Academy but across the state's 132 school divisions; and

4. Digital, on-demand classroom instruction in STEAM subjects and interdisciplinary studies for any student anywhere.

To advance its mission, the Virginia STEAM Academy's Board of Directors established five committees: (1) Governance; (2) Fund Development; (3) Teaching and Learning; (4) Human Capital; and (5) Buildings, Grounds, and Site Configuration. For our purposes here, we focus on highlights of the Teaching and Learning committee, although all five committees continue to work aggressively, simultaneously, and corporately to advance the full implementation of the Virginia STEAM Academy. Although the centrality of teaching and learning is not to be debated, it must also be recognized that the infrastructure of governance processes, the availability of resources, and a uniquely adequate space for the school are necessary for providing the context for a living-learning community, such as that envisioned for the Virginia STEAM Academy.

The Academy is based on a consortium leadership model. From its earliest days, Virginia STEAM's cofounders determined they would include a broad array of stakeholders in the Academy's design, delivery, measurement, and continuous improvement. Within the first years of planning and development, the Virginia STEAM Academy had signed partnership agreements with six universities, five science and engineering entities, one health system, and one arts institution. This commitment to inclusion has produced both challenge and opportunity. The challenge is obvious: Bringing multiple parties to the table can at times create tension, especially among parties who do not necessarily speak the same jargon or agree on the same intended educational or economic outcomes of a project of this scope and nature. But therein lies the opportunity. Listening to diverse perspectives across regions, populations, work, and disciplines, and intentionally acting upon those perspectives when they align with a core set of Virginia STEAM Academy principles, has helped establish a foundation for "real-world" partnerships, a defining characteristic of effective STEM-focused schools (Lynch, Peters-Burton, & Ford, 2014/2015). Figure 15.1 presents a conceptual model of the leadership consortium for the Virginia STEAM Academy.

Poised for Innovation

The Virginia STEAM Academy asserts that for students to be internationally competitive, instructional content must align to nationally and internationally benchmarked learning and performance standards (Crean Davis, 2013). Such alignment is desirable, but standardization to the point of restraining innovation is not (Sahlberg, 2006). Similarly, the Academy's approach to curriculum and instruction does not dismiss equally important areas of study within the humanities and social sciences. The Virginia STEAM Academy will couch the rigor of STEM disciplines within a learning environment that honors all disciplines. The Academy culture will encourage students, faculty, and partners to strive not only for mastery of what is "known," but to pursue what is unknown. For example, one of the primary intended outcomes for student learning will be the development of the appreciation for and capacity to engage in inquiry through multiple disciplinary lenses (National Research Council, 2012; NGSS Lead States, 2013).

. . . one of the primary intended outcomes for student learning will be the development of the appreciation for and capacity to engage in inquiry through multiple disciplinary lenses (National Research Council, 2012; NGSS Lead States, 2013).

Based on international comparison data, Sahlberg (2006) posited that many nations with high-performing education systems are moving away from sole demonstration of core knowledge and skill mastery as evidence of excellence. Instead, they are emphasizing deep knowledge and skill mastery—plus flexibility, creativity, and problem solving (Partnership for 21st Century Learning, 2011). The shift from what teachers teach to what students learn and do is in response not only to economic and social

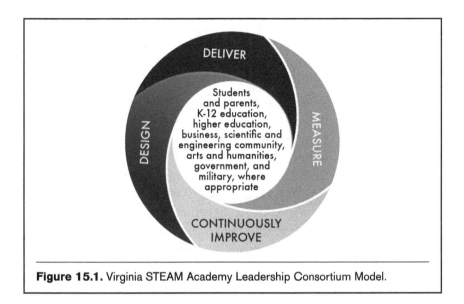

Figure 15.1. Virginia STEAM Academy Leadership Consortium Model.

demands, but also to understandings of how humans learn best (NGSS Lead States, 2013).

An inquiry-based and innovation-rich environment depends on collective intelligence, shared knowledge, and team-based problem solving (Hargreaves, 2003). By fostering an atmosphere for inquiry and innovation, the Academy aims to contribute to the increased economic competitiveness of the Commonwealth of Virginia and the country (United States Department of Commerce, 2012). More immediately and perhaps more importantly, the Academy envisions the nurturing and equipping of new generations of STEM leaders whose innate interests, talents, and dispositions are richly developed (Wagner, 2012).

A Distinctive Curriculum

A hallmark of effective curriculum design and implementation is the presence of a guaranteed and viable curriculum—that is, a clear set of experiences and intended learning outcomes that each and every student in the school has the opportunity to engage with and acquire (Marzano, 2003). To that end, the Virginia STEAM Academy will provide a rigorous, internationally benchmarked and 21st-century-relevant curriculum that tightly integrates STEM content, applied learning, the humanities, leadership, and ethics (Crean Davis, 2013; Fleischman, Hopstock, Pelczar, & Shelley, 2010; Vasquez, 2014/2015).

Broadly speaking, the program of studies will be familiar to many magnet, college preparatory, and STEM schools: advanced and college level courses; core and elective classes in science, technology, engineering, and mathematics; as well as courses in wellness (health/physical education), social sciences, English, world languages, and fine arts. Additionally, there will be opportunities for interdisciplinary study, entrepreneurship experiences, creative individual study, and research. While earning a high school diploma and pursuing advanced college credits, students will have the opportunity to obtain industrial certification; routinely collaborate with their peers, faculty, and field experts to address real-world challenges; experience sustained apprenticeships with partner organizations in STEAM-related occupations; and produce culminating projects that demonstrate STEAM-related outcomes (Lynch, Peters-Burton, & Ford, 2014/2015).

> . . . the Virginia STEAM Academy will provide a rigorous, internationally benchmarked and 21st-century-relevant curriculum that tightly integrates STEM content, applied learning, the humanities, leadership, and ethics.

It is worth calling attention to the Virginia STEAM Academy's commitment to a comprehensive curriculum. Researchers suggest a certain degree of mathematical maturity is required to fully participate in a 21st-century economy (Peterson, Woessmann, Hanushek, & Lastra-Anadón, 2011). Some might think, then, that a STEM-focused academy is a STEM-exclusive academy. Virginia STEAM Academy does not devalue the humanities; it esteems them. In fact, Virginia STEAM Academy leadership holds that those students who appreciate and integrate STEM disciplines with the humanities will be the ones to thrive in an increasingly competitive global society. Late Apple CEO and cofounder Steve Jobs exemplified this idea when he said, "I read something that one of my heroes, Edwin Land of Polaroid, said about the importance of people who stand at the *intersection of humanities and sciences* [italics added], and I decided that's what I wanted to do" (Isaacson, W., 2011, p. xix).

In the fall of 2012, the Virginia STEAM Academy Teaching and Learning Committee convened a Standards-Setting Panel made up of experts from public K–12, higher education, scientific and engineering research and development, and business to review and provide feedback on student learning outcomes and the best available student curriculum standards (Seremet & Stack, 2013). Among the emerging and foundational themes, the Standards-Setting Panel recommended the following key principles to guide the further development and design of the curriculum.

First, *accept and use the national content standards to lead the Virginia STEAM Academy curriculum design.* There was universal consensus among the Standards-Setting Panelists that the national content standards, including Next Generation Science Standards (NGSS Lead States, 2013), provide a rigorous and cognitively demanding framework for the curriculum design of the Virginia STEAM Academy. Because these standards have been vetted by national experts and include processes of international benchmarking and alignment with college and university entry level courses, the panelists advised accepting them and using them for the next steps in designing the curriculum and instructional program. Curriculum development will include further and ongoing review of the national standards and Academy curricula by representatives of the educational, business, and science and engineering research communities. Student learning outcome measures are being developed in tandem with the curriculum.

Some might think, then, that a STEM-focused academy is a STEM-exclusive academy. Virginia STEAM Academy does not devalue the humanities; it esteems them. In fact, Virginia STEAM Academy leadership holds that those students who appreciate and integrate STEM disciplines with the humanities will be the ones to thrive in an increasingly competitive global society.

Second, *take a global view of the world to inform students' learning and their future work world.* The Standards-Setting Panel was universal and passionate about the importance of taking a global view in several aspects of the Virginia STEAM Academy curriculum design. Panelists noted the economic, social, and technological interconnectedness of our global society and that STEM disciplines exist within a larger, dynamic, international community. They noted the need for Virginia STEAM Academy students to (1) exhibit cross-cultural sensitivity and understanding; (2) develop the skills and abilities that bring about innovation, foster entrepreneurship, and promote positive change; and (3) appreciate and practice service to others. These have become three core intended learning outcomes of the Virginia STEAM Academy curriculum framework (Seremet & Stack, 2013).

Third, *give students opportunities to learn in a trans-disciplinary world.* The Standards-Setting Panel recommended that content disciplines be viewed as lenses through which multiple approaches toward learning are presented (Ornstein & Hunkins, 2013). A trans-disciplinary approach to learning engenders the habit of holding two opposing views in mind at once and making informed judgments. Several panelists provided specific cross-content linkages to assist faculty in tackling the complex task of teaching content and skills deeply in each subject area while allowing students to see and apply their learning across subject areas. Additionally, the curriculum will integrate opportunities for students to develop self-awareness and metacognitive understanding, such as about one's and others' learning styles; cognitive strengths; and social, emotional, and physical assets (Davies, Fidler, & Gobis, 2011). Relatedly, the Virginia STEAM Academy curriculum will provide students with opportunities to work with others and to be exposed to and explore possible career pathways.

Innovative Instructional Approach

As the means toward enacting the curriculum of the Virginia STEAM Academy, a number of key, innovative instructional approaches will characterize the school. The Standards-Setting Panel of the Teaching and Learning Committee identified the following instructional design principles.

First, *design learning experiences as "problem-based."* Whether through capstone experiences; internships with scientists, engineers, researchers, and/or business leaders; or classroom activities and assignments, panelists suggested that allowing time for projects that pull from several disciplines, where appropriate, should be designed into day-to-day schedules. Giving students learning opportunities to solve problems is engaging and highly motivating for adolescent learners. This is one of the core benefits and opportunities from a residential setting: Teaching and learning can occur "24/7" both within and outside the classroom; in teams of peers, students, and faculty; and applied in authentic settings with science, engineering, and business leaders.

As a second instructional design principle, *make "applied" pertinent to all the letters of STEM and to the humanities.* Panelists noted that the Virginia STEAM Academy leaders have selected a name with varying connotations in the field. Some will think the "A" means the arts. Others will assume the *applied* is referring only to the mathematics that follows it in the title. The Virginia STEAM Academy must clarify and communicate that the vision includes making the *applied* pertinent to science, technology, engineering, mathematics, and indeed, all of the subject areas. Seeking opportunities for students to build habits of mind in self-awareness, self-directed learning, and applying their skills to real-world challenges will require students to draw between and across their subject area lessons. Incorporating these opportunities will be an important component of the instructional design of the Academy (Davies, Fidler & Gorbis, 2011).

Third, *utilize the uniqueness of "statewide" and "residential."* Panelists spoke to the Virginia STEAM Academy's unique capacity to bring together a socio-economically, ethnically, and geographically diverse student population in a residential setting (Lynch, Peters-Burton, & Ford, 2014/2015). Highlighting the one, statewide site for student benefit and the educational laboratory setting in a STEAM-rich corridor with its multiple military, business, science, engineering, and research sites for student engagement and internships, will assist in recruiting both students and faculty. The scheduling options in a residential setting maximize how clubs and cocurricular activities will enhance student experiences. Mentoring opportunities for students to work with leaders in STEM fields and a chance for Virginia STEAM Academy students to become STEAM Ambassadors in their home school districts is both unique and important in what the Academy offers to its students. Compelling data from the North Carolina School of Science and Mathematics show that 60% of its graduates choose to live and work in North Carolina, thus bolstering the state's STEAM talent and economy (Karen Dash Consulting, 2011).

Fourth, *include intended learning outcomes that nurture ethical service and leadership*. A defining element of the mission of the Virginia STEAM Academy is its explicit articulation of the dispositions of ethical service and leadership as intended learning outcomes for students (Fisher & Frey, 2014/2015). To that end, the instructional program will seek to include student-selected community service and social justice experiences that empower students to make a difference in local issues. Learning will also take place through mentoring experiences where junior- and senior-level students mentor younger Academy students and/or middle and elementary students across the Commonwealth. Similarly, students will participate in long-term mentoring experiences with practicing engineers, scientists, researchers, and higher education faculty. A program with the intent to reinforce connections for students to their home school divisions and communities has been conceived, whereby students will have opportunities to serve as "STEAM Ambassadors" by returning to their home school districts as the face of the Virginia STEAM Academy. The global leadership outcomes for students to appreciate the needs and rewards of a lifetime of service will be integral to the instructional program. Meeting world challenges may counter the American cultural norm of immediate rewards, which will require the Academy staff to nurture patience and the long view and to extend the desired skills of grasping complex ideas and demonstrating perseverance despite obstacles.

Finally, the instructional program will be designed to *nurture the affective domain*. Virginia STEAM Academy faculty will need to have a depth of understanding of adolescent development, as well as of gifted education. Their professional development will regularly focus on issues of adolescent development as well as discipline-centered pedagogical content knowledge. Research shows gifted students often demonstrate advanced intellectual levels at the same time

A program with the intent to reinforce connections for students to their home school divisions and communities has been conceived, whereby students will have opportunities to serve as "STEAM Ambassadors" by returning to their home school districts as the face of the Virginia STEAM Academy.

they demonstrate weak physical and/or emotional levels (Silverman, 2013). For students at the Virginia STEAM Academy, the social-emotional processes for approaching and working through experiences will be as important a learning objective as the application of content knowledge and skills to specific tasks. Specifically, throughout their academic and residential experiences, Virginia STEAM Academy faculty and staff will actively reference, model, practice, and reflect upon habits of mind to shape learning, decision making, relationship building, and leadership qualities (Costa & Kallick, 2008). Internalizing this strategy will not only provide a useful process for approaching new experiences, but it will also create a setting in which students can confidently take risks, tackle complicated challenges, adapt to emerging circumstances, and serve as effective stewards of their intellectual gifts. This approach is consistent with leadership practices that are valued and exemplified by highly respected CEOs throughout the world (IBM, 2012). It is also a hallmark of some of the nation's most esteemed private boarding schools (Cookson & Persell, 1987).

Impact

Although the boarding high school has yet to be launched, the Virginia STEAM Academy has offered three highly successful Summer STEAM academies intended to inspire and expose rising sixth through eighth graders from across the Commonwealth to STEM content, applied learning, and leadership development. An independent evaluation found:

- ▷ 98% of parents said they would recommend Summer STEAM;
- ▷ 94.5% of students reported learning new things;
- ▷ 91% reported improvement of mathematical, scientific, and engineering skills;
- ▷ 87% reported learning to respect opinions that are different;
- ▷ 80% reported developing leadership skills; and
- ▷ 80% reported improvement in their ability to work with a group to accomplish a task (Hobson, 2014).

To date, the Virginia STEAM Academy has 187 Summer STEAM Ambassadors representing every region of the Commonwealth. Summer STEAM Ambassadors are among Virginia's most able and interested STEM students. Summer STEAM is offered as a public-private program to Virginia's young scholars at no cost to parents. Thus, for many young scholars, Summer STEAM is their only opportunity to experience the rigor and relationships that evolve out of an immersive living-learning laboratory with like-minded, like-motivated peers, and exceptional faculty and staff. Summer STEAM students were selected from close to 700 applicants to participate in the weeklong, summer learning experience. In the program, students spend approximately 3 hours each day in content-specific learning labs, followed by lunch and social time, and then another 3 hours in applied learning labs. This daily integration of content and applied learning has a positive ripple effect, namely highlighting that learning does not end with the official learning lab. Highly able and motivated students learn with their peers before, during, and after class. They continue to ask probing questions of their faculty, teaching assistants, and resident assistants. They grapple with challenges "off hours" because they want to do so. And they enjoy the down time of play to experiment, recharge, and build relationships (Wagner, 2012).

> Highly able and motivated students learn with their peers before, during, and after class. They continue to ask probing questions of their faculty, teaching assistants, and resident assistants. They grapple with challenges "off hours" because they want to do so. And they enjoy the down time of play to experiment, recharge, and build relationships (Wagner, 2012).

Here's one example of a recent student's experience. Students at the Virginia STEAM Academy's inaugural Summer STEAM session could choose between two tracks: math reasoning and encryption, and physics. Students who participated in math reasoning and encryption studied advanced problem-solving techniques in the morning, including RSA code, and applied it to a code-breaking lab in the afternoon. The encryption teacher was an experienced research scientist (holding a doctorate) with Virginia Modeling, Analysis, and Simulation Center, a research division of Old Dominion University. He presented the students with a "throw-away problem." He told them a particular code caused a computer virus that no one had heretofore solved. Once infected, the computer repeats 0-1-0-1-0-1 until it is shut down. He went on with the lesson and the students participated enthusiastically. After the instructor adjourned class, the students hopped on the bus to return to the host university campus. On the bus ride back to campus, a small group of students sat together and solved the "unsolvable" virus.

Such highly motivated, inquisitive, and exceptionally able students deserve a specialized education where the conditions are ripe for immersive, rigorous, contextualized learning with similarly able and motivated peers, expert faculty, and practitioners (Finn & Hockett, 2012). The Virginia STEAM Academy provides that environment for such students and for the families who desire a 24/7 immersive, living-learning experience.

The success of the three Summer STEAM academies has served as a promising indicator of the anticipated success of the residential Virginia STEAM Academy model. Lessons learned from Summer STEAM and from study and dialogue with colleagues in government, education, science, engineering, industry, business, military, and the arts have increased the groundswell of support for the school. State and private funds have been secured to continue the development and design of the Virginia STEAM Academy curriculum and applied learning experiences. Partnerships are being expanded and new ones formalized.

The school has also identified a preferred home. The Virginia STEAM Academy secured a Memorandum of Understanding with Fort Monroe in Hampton, VA, as the preferred campus site for the residential school. This is an exciting prospect, as Fort Monroe is state-owned property, is of historic significance, and sits at the center of the Hampton Roads region of southeastern Virginia, which is a center of corporate, industry, and military activity. Building and site renovation projections, timeline and costs, and one-line drawings of all buildings have been created.

The Virginia STEAM Academy has also been admitted as an Associate Member of the National Consortium of Secondary STEM Schools (NCSSS), a 100-plus member consortium of secondary schools and 75 affiliated members representing nearly 40,000 students and 1,600 educators committed to preparing students for leadership in science, technology, engineering, and mathematics.

The Virginia STEAM Academy's success will mark a milestone in the Commonwealth's education history and economic future. For the first time, access to world-class education and mentorship opportunities with secondary teachers, university faculty, and practicing scientists and engineers will be available to any qualifying high school student regardless of zip code.

> Such highly motivated, inquisitive, and exceptionally able students deserve a specialized education where the conditions are ripe for immersive, rigorous, contextualized learning with similarly able and motivated peers, expert faculty, and practitioners (Finn & Hockett, 2012). The Virginia STEAM Academy provides that environment for such students and for the families who desire a 24/7 immersive, living-learning experience.

Next Steps

The Virginia STEAM Academy Board of Directors will make a careful assessment of its progress to date and announce a timetable for opening the boarding high school. The Academy will offer its third year of summer residential academies for approximately 100 middle school students in summer 2015.

Among its other next steps, the Virginia STEAM Academy will expand its board of directors, hire leadership positions to complete preparation to welcome the Virginia STEAM Academy's first class, and prepare appropriate legislative language and budgets for pre-opening, phased capital build out, and private and public funding for sustainable operations. It will draft institutional policies and procedures, develop curriculum and more specific courses and assessments with lead teachers and pursue accreditation through AdvancEd. Finally, it will secure its lease agreement and renovate essential buildings.

As professionals and concerned citizens, we must look at how we serve all students, including our exceptionally able. It is tempting but false to assume these students will thrive simply because of their intellectual ability. They, too, must be nurtured. They, too, must be provided a learning environment that is guided by national standards and effective practice, and that nurtures their full development.

Discussion Questions

1. What curricular components and plans for instructional vision are included in the design of the Virginia STEM Academy?
2. In planning a STEM program or school for talented students, which groups of stakeholders should be involved at each of the different stages of program or school development?

References

Carnevale, A. P., Smith, N., & Strohl, J. (2010). *Help wanted: Projections of jobs and education requirements through 2018*. Washington, DC: Georgetown University Center on Education and the Workforce.

Change the Equation. (2014). Vital signs: Virginia. Retrieved from http://vitalsigns.changetheequation.org/tcpdf/vitalsigns/newsletter.php?statename=Virginia

Cookson, P. W., & Persell, C. H. (1987). *Preparing for power: America's elite boarding schools.* New York, NY: Basic Books.

Costa, A. L., & Kallick, B. (Eds.) (2008). *Learning and leading with habits of mind.* Alexandria, VA: Association for Supervision and Curriculum Development.

Crean Davis, A. (2013). *Beyond our borders: The value of international benchmarks for the Virginia STEAM Academy.* A commissioned paper for the Virginia STEAM Academy. Suffolk, VA: Virginia STEAM Academy.

Davies, A., Fidler, D., & Gorbis, M. (2011). *Future work skills 2020.* Palo Alto, CA: Institute for the Future.

Finn, C. E., & Hockett, J. A. (2012). *Exam school: Inside America's most selective public high schools.* Princeton, NJ: Princeton University Press.

Fisher, D., & Frey, N. (2014/2015). STEM for citizenship. *Educational Leadership 72*(4), 86–87.

Fleischman, H. L., Hopstock, P. J., Pelczar, M. P., & Shelley, B. E. (2010). *Highlights from PISA 2009: Performance of U.S. 15-year-old students in reading, mathematics, and science literacy in an international context* (NCES 2011–04). Washington, DC: U.S. Department of Education, National Center for Education Statistics.

Hargreaves, A. (2003). *Teaching in the knowledge society: Education in the age of insecurity.* New York, NY: Teachers College Press.

Hobson, E. (2014). Program evaluation of summer STEAM 2014. Report prepared for the Virginia STEAM Academy. Hampton, VA: Virginia STEAM Academy.

IBM. (2012). *Leading through connections: Insights from the global chief executive officer study.* Somers, NY: Author.

Isaacson, W. (2011). *Steve Jobs.* New York, NY: Simon & Schuster.

Karen Dash Consulting. (2011). Economic impact statement. Report prepared for the North Carolina School of Science and Mathematics. Durham, NC: Author.

Kouzes, J. M., & Posner, B. Z. (2002). *The leadership challenge* (3rd ed.). San Francisco, CA: Jossey-Bass.

Lederman, L. M. (1998). *ARISE: American renaissance in science education.* Batavia, IL: Fermi National Accelerator Laboratory.

Lynch, S. J., Peters-Burton, E., & Ford, M. (2014/2015). Building STEM opportunities for all. *Educational Leadership 72*(4), 54–60.

Marzano, R. J. (2003). *What works in schools: Translating research into action.* Alexandria, VA: Association for Supervision and Curriculum Development.

National Research Council. (2012). *A framework for K–12 science education: Practices, crosscutting concepts, and core ideas.* Washington, DC: National Academies Press.

NGSS Lead States. (2013). *Next Generation Science Standards: For states, by states.* Washington, DC: National Academies Press.

Ornstein, A. C., & Hunkins, F. P. (2013). *Curriculum foundations, principles, and issues* (6th ed.). Boston, MA: Pearson.

Partnership for 21st Century Learning. (2011). *Framework for 21st century learning.* Retrieved from http://www.p21.org/our-work/p21-framework

Peterson, P. E., Woessmann, L., Hanushek, E. A., & Lastra-Anadón, C. X. (2011, Aug.). *Globally challenged: Are U.S. students ready to compete? The latest on each state's international standing in math and reading.* PEPG Report No. 11-03. Cambridge, MA: Harvard's Program on Education Policy and Governance and Education.

Sahlberg, P. (2006). Education reform for raising economic competitiveness. *Journal of Educational Change 7*(4), 259–287.

Seremet, C. P., & Stack, D. L. (2013). *Building from the Standards-Setting Panel to curriculum and instructional design: Setting the course for the Virginia STEAM Academy. A commissioned paper for the Virginia STEAM Academy.* Suffolk, VA: Virginia STEAM Academy.

Silverman, L. K. (2013). Asynchronous development: Theoretical bases and content applications. In C. S. Neville, M. M. Piechowski, & S. S. Tolan (Eds.), *Off the charts: Asynchrony and the gifted child.* Unionville, NY: Royal Fireworks Press.

United States Department of Commerce. (2012). *The competitiveness and innovative capacity of the United States.* Washington, DC: Author.

Vasquez, J. A. (2014/2015). STEM: Beyond the acronym. *Educational Leadership 72*(4), 10–15.

Virginia Chamber of Commerce (2013). *Blueprint Virginia: A business plan for the Commonwealth.* Richmond, VA: Author.

Wagner, T. (2012). *Creating innovators: The making of young people who will change the world.* New York, NY: Scribner.

Conclusion

This book has been developed to fill an important need for discussing and supporting differentiated learning for high-ability students in STEM educational settings. *STEM Education for High-Ability Learners: Designing and Implementing Programming* is a professional resource for educators, researchers, and parents planning comprehensive learning experiences for enhancing high-ability students' understandings about the varied dimensions of STEM.

Educators who work with advanced students are generally known for their work in teaching excellence. Talented teachers have confidence that their students will go into the world and do great things. As people move, distances increase and we may not know what is happening in the lives of each former student. Sometimes students' stories may return to teachers, and it is exciting to hear of students' successes. Teaching provides great satisfaction and also delayed gratification. Cultivating technical talent is a long and winding process and how teachers contribute to shaping the journey is what makes the teaching profession so rewarding. I hope you enjoy this final student story from an unexpected place related to STEM talent development:

Austin, TX—After a capitol tour and while closely examining a framed historical document on the wall in the Texas House of Representatives, a voice behind me asked, "Do you remember me?" I turned and saw a young man with his wife watching me. It had been 10 years since I left my classroom in a rural school district to pursue my doctoral studies, where I shared with my young students that I would be going to the oldest school in Virginia to study for my doctorate. Much has transpired in the decade that has passed since, and periodically I hear from former students who stay in touch with news to share. The quiet young man standing before me in the Texas state house was in my gifted classroom in rural Missouri. He shared that he was studying for his doctorate at Texas A&M in College Station in materials science engineering (one of the areas which my father specialized). We chatted a while about life and he smiled as I asked about his parents, recognizing that indeed he was remembered well. His young wife excitedly told on him exclaiming that, "He kept saying, that's her, that's her, I know it!!" We laughed together and enjoyed how serendipitous it was to meet in this place. As we walked away, my new husband beamed at me, and as we drove down a street in Austin, we passed a billboard that read, "Don't worry about being famous, if you want to be remembered, Teach."

What a blessing it is when teachers can come full circle and reconnect with former students. And it always seems to happen at random times, as if when needed the most. The impact teachers have on students is extraordinary, as teachers work daily stirring and flaming bright intellectual sparks in their pupils. The impact a teacher can make upon a student is profound indeed, and as my former principal used to say, "Through this school walks the best students, faculty, and staff in the world." I hope you enjoy shaping and sharing in students' success stories of their technical talent development.

About the Editor

Bronwyn MacFarlane, Ph.D., is associate professor of gifted education at the University of Arkansas at Little Rock (UALR) in the Department of Educational Leadership. She has served as interim Associate Dean of the UALR College of Education and Health Professions providing leadership for educator and health preparation programs. She earned her doctorate in Educational Leadership, Policy, and Planning from the College of William and Mary. Dr. MacFarlane co-edited *Leading Change in Gifted Education: The Festschrift of Dr. Joyce VanTassel-Baska* (2009) and her research interests focus upon talent development, educational programming, and curricula interventions.

About the Authors

Scott Chamberlin, Ph.D., is an associate professor at the University of Wyoming in the field of mathematics education. His research interests include the use of problem-solving activities with upper elementary and middle grade gifted students. Specifically, he investigates student affect and creativity in relation to mathematical problem solving. He developed the Chamberlin Affective Instrument for Mathematical Problem Solving (CAIMPS), and much of this work is a direct result of his work with Model-eliciting Activities (MEAs). MEAs are problem-solving tasks in which solvers are expected to create mathematical models to explain and generate understanding about phenomena and concepts in mathematics.

Alicia Cotabish, Ed.D., is Gifted Education Program Coordinator at the University of Central Arkansas. Previously, she served as one of two Principal Investigators of STEM Starters and the program coordinator of the Arkansas Evaluation Initiative in Gifted Education (AEI), both federally funded Jacob K. Javits projects housed at the University of Arkansas at Little Rock. She is president of the Arkansas Association of Gifted Education Administrators. Her recent research has focused on K–20 STEM and gifted education and examining the effects of virtual coaching on the quality of teacher candidates using Skype and Bluetooth Bug-in-the-Ear (BIE) technology.

Steve Coxon, Ph.D., is an associate professor and Director of Programs in Gifted Education at Maryville University, including the gifted education graduate program; the Maryville Young Scholars Program to increase diversity in gifted programs; the Children using Robotics for Engineering, Science, Technology, and Math (CREST-M) project to create math curricula to engage and prepare diverse students for STEM careers; and the Maryville Summer Science and Robotics Program. Steve conducts research on developing STEM talents, is author of numerous publications including the book *Serving Visual-Spatial Learners*, is the science education columnist for *Teaching for High Potential*, and book review editor for *Roeper Review*.

Debbie Dailey, Ed.D., is an assistant professor of teaching and learning at the University of Central Arkansas, where she instructs courses in the Master of Arts in Teaching program and the Gifted and Talented Education program. Formerly, Debbie was the Associate Director for the Jodie Mahony Center for Gifted Education and Advanced Placement at the University of Arkansas at Little Rock. Debbie also served as the Curriculum Coordinator and Peer Coach of a federally funded program, STEM Starters, which focused on improving science instruction in the elementary grades. Prior to moving to higher education, Debbie was a high school science teacher and gifted education teacher for 20 years.

Mary Christine Deitz, Ed.D., is the specialist for gifted and talented programs for the Little Rock School District. She was recognized as the Doctoral Student of the Year by the National Association of Gifted Children (NAGC) and received the A. Harry Passow Classroom Teacher Award, Graduate Student of the Year, and Curriculum Award from NAGC. Dr. Deitz has taught in public schools for more than 26 years and worked exclusively with diverse populations of high-ability children. Known for developing innovative and engaging instruction for high-ability learners, Dr. Deitz is a consultant for district, state, and national educational organizations.

Christopher R. Gareis, Ed.D., is an associate professor of educational leadership in the School of Education at the College of William and Mary. He has twice served as Associate Dean for Teacher Education. His expertise is in instructional leadership and classroom assessment, and he has served on various subcommittees for the Virginia STEAM Academy.

Angela M. Housand, Ph.D., is an associate professor at the University of North Carolina Wilmington and a national consultant. As a former teacher, Dr. Housand brings an applied focus to her instructional programs for teachers, as well as her research testing the effectiveness of the FutureCasting digital life skills program. Over the years, her work has been presented internationally and published in top journals. Her efforts directly support teachers as they challenge students to achieve advanced levels of performance while becoming productive citizens in a global society. For more, visit http://www.angelahousand.com.

Brian C. Housand, Ph.D., is an associate professor and co-coordinator of the Academically and Intellectually Gifted Program at East Carolina University. In 2014, he received the Max Ray Joyner Award for Outstanding Teaching in Distance Education at ECU. Dr. Housand earned a Ph.D. in educational psychology at the University of Connecticut's Neag Center for Gifted Education and Talent Development with an emphasis in both gifted education and instructional technology. He serves on the National Association for Gifted Children Board of Directors as a Member-at-Large. He researches ways in which technology can enhance the learning environment and is striving to define creative-productive giftedness in a digital age. His website is http://www.brianhousand.com.

Barbara Kerr, Ph.D., holds an endowed chair as Distinguished Professor of Counseling Psychology at the University of Kansas and is an American Psychological Association Fellow. Her M.A. from Ohio State University and her Ph.D. from the University of Missouri are both in counseling psychology. She directs the Counseling Laboratory for the Exploration of Optimal States (CLEOS), serving creative adolescents and adults. She is author of *Smart Girls in the Twenty-First Century* and coauthor of *Smart Boys: Talent, Masculinity, and the Search for Meaning, Counseling Girls and Women,* and more than 100 articles, chapters, and papers in the areas of talent, creativity, and gender issues.

Kristy Kidd, M.Ed., is the Project Director of STEM Starters Plus, a federally funded Javits project at the University of Arkansas at Little Rock and the Jodie Mahony Center for Gifted Education. She has more than 21 years of experience teaching elementary and middle school science in Little Rock Public Schools. She has served as the K–12 Math and Science Specialist for the eStem Public Charter Schools and as an adjunct instructor for early childhood science methods courses at the University of Arkansas at Little Rock. She received the Milken National Educator Award, which honors top educators across the country. She is also a state finalist for the Presidential Award for Excellence in Mathematics and Science Teaching.

Eric Mann, Ph.D., is an assistant professor of mathematics education at Hope College in Holland, MI. After completing an Air Force career, he taught math and science for 7 years before entering the doctoral program in educational psychology at the University of Connecticut's Neag Center for Gifted Education. Dr. Mann served on the faculty at Purdue University's Institute for P–12 Engineering Research and Learning and the Gifted Education Resource

Institute before accepting his current position. He is interested in a deeper understating of creativity and talent development within science, technology, mathematics, and engineering (STEM) disciplines, with an emphasis on STEM literacy and engagement.

M. Caroline Martin, RN, MHA, is cofounder and Board President of the Virginia STEAM Academy. She is the retired Executive Riverside Health System CEO at the Riverside Regional Medical Center.

Rachelle Miller received her Ph.D. in educational psychology and is an assistant professor in the Department of Teaching and Learning at the University of Central Arkansas. She works in the Gifted and Talented Education program teaching Affective Strategies for the Gifted and Talented. She currently collaborates with Arkansas A+ Schools by completing program evaluation and assisting in the development of arts integrated curriculum in their participating schools. Her research interests include supporting the academic needs of low-income gifted students, integrating the arts into the general and gifted curriculum, and examining teacher perceptions of arts integration.

Heather A. Olvey is a graduate student at the University of Arkansas at Little Rock. As a candidate in the master's program in Secondary English Education, she also serves as the graduate assistant for the UALR Teacher Education Department. She has coauthored two book chapters and several articles about the role of young adult literature and how to use it in secondary classrooms. She has assisted in qualitative research projects examining the effects of young adult literature upon changing student and teacher perceptions about social justice issues, and has presented the research at the International Reading Association (IRA) and other regional organizations.

Julia L. Roberts, Ed.D., Mahurin Professor of Gifted Studies at Western Kentucky University, is Executive Director of The Carol Martin Gatton Academy of Mathematics and Science in Kentucky and The Center for Gifted Studies. Dr. Roberts is on the Executive Committee of the World Council for Gifted and Talented Children and past-president of The Association for the Gifted. Her writing focuses on differentiation, gifted education, STEM, and advocacy. She received the 2011 Acorn Award as the outstanding professor at a Kentucky 4-year university, the first NAGC David Belin Advocacy Award, and the 2011 William T. Nallia Award for innovative leadership from the Kentucky Association for School Administrators.

Ann Robinson, Ph.D., is a professor of educational psychology and Founding Director of the Jodie Mahony Center for Gifted Education at the University of Arkansas at Little Rock. She is past-president of the National Association for Gifted Children (NAGC), a former editor of *Gifted Child*

Quarterly, and has been honored by NAGC as Early Scholar, Early Leader, Distinguished Scholar, and for Distinguished Service to the association. Her interests include the use of biography in curriculum, biographical research methods in gifted education, school intervention studies, evidence-based practices, and teacher preparation and professional development. She is the lead author on the best-selling *Best Practices in Gifted Education: An Evidence-Based Guide.* Her most recent book, co-edited with Jennifer Jolly, is *A Century of Contributions to Gifted Education: Illuminating Lives.*

Amy Sedivy-Benton, Ph.D., is an assistant professor in the Teacher Education Department at the University of Arkansas at Little Rock. She received her doctorate in research methodology from Loyola University of Chicago. Her research methodological expertise ranges from qualitative analysis to advanced statistical techniques. Dr. Sedivy-Benton works extensively with students in their continuing education to become teachers and administrators. Key areas of her work focus on mathematics and science content areas, difficult-to-staff schools, and the use of advanced statistics in concert with how these methods and structures fit into changes in educational policy. She has presented both nationally and regionally and serves on several journals as a consulting editor.

Judy K. Stewart, Ph.D., is cofounder, board member, and President and CEO of the Virginia STEAM Academy. She worked for several years in the educational research and development field, including owning a consulting business.

James Van Haneghan, Ph.D., is a professor of professional studies in the College of Education at the University of South Alabama, where he teaches courses in research methods, assessment, and learning. He has research interests in the areas of evaluation, problem- and project-based learning, mathematics education, motivation, and assessment. Before moving to South Alabama, he held positions at Northern Illinois University and George Peabody College of Vanderbilt University. His doctoral training was at the Applied Developmental Psychology Program at the University of Maryland. He also holds an M.A. in experimental psychology from S.U.N.Y at Geneseo, and a B.S. from S.U.N.Y.at Brockport.

J. D. Wright is a doctoral student in the University of Kansas Counseling Psychology program. He previously served as an adjunct instructor at the University of Florida in the School of Human Development and Organizational Studies in Education. He currently works with creative and gifted students through the Counseling Laboratory for the Exploration of Optimal States (CLEOS), offering targeted career counseling for high-achieving youths. He also utilizes EEG neurofeedback combined with guided visualizations to teach the students about creativity and flow states.

CPSIA information can be obtained
at www.ICGtesting.com
Printed in the USA
LVHW011511290423
745601LV00005B/453

9 781618 214324